DYNAMICS of SECOND ORDER RATIONAL DIFFERENCE EQUATIONS

With Open Problems and Conjectures

M.R.S. Kulenović
G. Ladas

CHAPMAN & HALL/CRC

Boca Raton London New York Washington, D.C.

Library of Congress Cataloging-in-Publication Data

Kulenović, M.R.S. (Mustafa R.S.)
Dynamics of second order rational difference equations : with open problems and conjectures / M.R.S. Kulenović, G. Ladas.
p. cm.
Includes bibliographical references and index.
ISBN 1-58488-275-1
1. Difference equations—Numerical solutions. I. Ladas, G.E. II. Title.
QA431 .K84 2001
515′.625—dc21 20010370

Direct all inquiries to CRC Press LLC, 2000 N.W. Corporate Blvd., Boca Raton, Florida 33431.

Visit the CRC Press Web site at www.crcpress.com

International Standard Book Number 1-58488-275-1
Library of Congress Card Number 20010370
Printed in the United States of America 1 2 3 4 5 6 7 8 9 0
Printed on acid-free paper

To

Senada and Theodora

Contents

Preface

In this monograph, we present the known results and derive several new ones on the boundedness, the global stability, and the periodicity of solutions of all rational difference equations of the form

$$x_{n+1} = \frac{\alpha + \beta x_n + \gamma x_{n-1}}{A + Bx_n + Cx_{n-1}}, \quad n = 0, 1, \ldots \tag{1}$$

where the parameters $\alpha, \beta, \gamma, A, B, C$ and the initial conditions x_{-1} and x_0 are nonnegative real numbers.

We believe that the results about Eq(1) are of paramount importance in their own right and furthermore we believe that these results offer prototypes towards the development of the basic theory of the global behavior of solutions of nonlinear difference equations of order greater than one. The techniques and results which we develop in this monograph to understand the dynamics of Eq(1) are also extremely useful in analyzing the equations in the mathematical models of various biological systems and other applications.

It is an amazing fact that Eq(1) contains, as special cases, a large number of equations whose dynamics have not been thoroughly understood yet and remain a great challenge for further investigation. To this end we pose several **Open Problems and Conjectures** which we believe will stimulate further interest towards a complete understanding of the dynamics of Eq(1) and its functional generalizations.

Chapter 1 contains some basic definitions and some general results which are used throughout the book. In this sense, this is a self-contained monograph and the main prerequisite that the reader needs to understand the material presented here and to be able to attack the open problems and conjectures is a good foundation in analysis. In an appendix, at the end of this monograph, we present some global attractivity results for higher order difference equations which may be useful for extensions and generalizations.

In Chapter 2 we present some general results about periodic solutions and invariant intervals of Eq(1). In particular we present necessary and sufficient conditions for Eq(1) to have period-two solutions and we also present a table of invariant intervals.

In Chapters 3 to 11 we present the known results and derive several new ones on the various types of equations which comprise Eq(1).

Every chapter in this monograph contains several open problems and conjectures related to the particular equation which is investigated and its generalizations.

Acknowledgements

This monograph is the outgrowth of lecture notes and seminars which were given at the University of Rhode Island during the last five years. We are thankful to Professors R. D. Driver, E. A. Grove, J. Hoag, W. Kosmala, L. F. Martins, O. Merino, S. Schultz, and W. S. Sizer and to our graduate students A. M. Amleh, W. Briden, E. Camouzis, C. A. Clark, K. Cunningham, R. C. DeVault, H. El-Metwally, J. Feuer, C. Gibbons, E. Janowski, C. Kent, L. McGrath, M. Predescu, N. Prokup, M. Radin, I. W. Rodrigues, C. T. Teixeira, S. Valicenti, and K. P. Wilkinson for their enthusiastic participation and useful suggestions which helped to improve the exposition.

Introduction and Classification of Equation Types

In this monograph, we present the known results and derive several new ones on the boundedness, the global stability, and the periodicity of solutions of all rational difference equations of the form

$$x_{n+1} = \frac{\alpha + \beta x_n + \gamma x_{n-1}}{A + Bx_n + Cx_{n-1}}, \quad n = 0, 1, \ldots \tag{1}$$

where the parameters $\alpha, \beta, \gamma, A, B, C$ are nonnegative real numbers and the initial conditions x_{-1} and x_0 are arbitrary nonnegative real numbers such that

$$A + Bx_n + Cx_{n-1} > 0 \quad \text{for all} \quad n \geq 0.$$

We believe that the results about Eq(1) are of paramount importance in their own right and furthermore we believe that these results offer prototypes towards the development of the basic theory of the global behavior of solutions of nonlinear difference equations of order greater than one. The techniques and results which we develop in this monograph to understand the dynamics of Eq(1) are also extremely useful in analyzing the equations in the mathematical models of various biological systems and other applications.

It is an amazing fact that Eq(1) contains as special cases, a large number of equations whose dynamics have not been thoroughly established yet. To this end we pose several **Open Problems and Conjectures** which we believe will stimulate further interest towards a complete understanding of the dynamics of Eq(1) and its functional generalizations.

Eq(1) includes among others the following well known classes of equations:

1. The **Riccati difference equation** when

$$\gamma = C = 0.$$

This equation is studied in many textbooks on difference equations, such as [1], [21], [41], [42], etc. See Section 1.6 where, in addition to the basic description of the solutions of the Riccati equation, we introduce the **forbidden set** of the equation. This is the set $\mathbf{F}$ of all initial conditions $x_0 \in \mathbf{R}$ through which the equation

$$x_{n+1} = \frac{\alpha + \beta x_n}{A + B x_n} \tag{2}$$

is not well defined for all $n \geq 0$. Hence the solution $\{x_n\}$ of Eq(2) exists for all $n \geq 0$, if and only if $x_0 \notin \mathbf{F}$.

The problem of **existence of solutions** for difference equations is of paramount importance but so far has been systematically neglected. See [13], [18], [32], [69], and Section 1.6 for some results in this regard.

Throughout this monograph we pose several open problems related to the existence of solutions of some simple equations with the hope that their investigation may throw some light into this very difficult problem.

An example of such a problem is the following:

Open Problem *Find all initial points $(x_{-1}, x_0) \in \mathbf{R} \times \mathbf{R}$ through which the equation*

$$x_{n+1} = -1 + \frac{x_{n-1}}{x_n}$$

is well defined for all $n \geq 0$. See [13].

2. **Pielou's discrete delay logistic model** when

$$\alpha = \gamma = B = 0.$$

See ([42], p. 75), [55], [66], [67], and Section 4.4.

We believe that the techniques which we develop in this monograph to understand the dynamics of Eq(1) will also be of paramount importance in analyzing the equations in the mathematical models of various biological systems and other applications. For example, it is interesting to note the similarities of the dynamics of the population model

$$x_{n+1} = \alpha + \beta x_{n-1} e^{-x_n}, \quad n = 0, 1, \ldots$$

and the rational equation

$$x_{n+1} = \frac{\alpha + \alpha x_n + \beta x_{n-1}}{1 + x_n}, \quad n = 0, 1, \ldots .$$

See [24] and Sections 2.7 and 10.2.

3. **Lyness' equation** when

$$\gamma = A = B = 0.$$

See [42], [44], [49], [57], and Section 5.2. This is a fascinating equation with many interesting extensions and generalizations. A characteristic feature of this equation is that it possesses an **invariant** which can be used to understand the character of its solutions. Without the use of this invariant we are still unable to show, for example, that every positive solution of the difference equation

$$x_{n+1} = \frac{\alpha + x_n}{x_{n-1}}, \quad n = 0, 1, \ldots$$

where $\alpha \in (0, \infty)$, is bounded.

It is an amazing fact that Eq(1) contains, as special cases, a large number of equations whose dynamics have not been thoroughly established yet.

For convenience we classify all the special cases of Eq(1) as follows.

Let k and l be positive integers from the set $\{1, 2, 3\}$. We will say that a special case of Eq(1) is of (k, l)-type if the equation is of the form of Eq(1) with k positive parameters in the numerator and l positive parameters in the denominator.

For example the following equations

$$x_{n+1} = \frac{x_n}{x_{n-1}}, \quad n = 0, 1, \ldots$$

$$x_{n+1} = \frac{\alpha + \beta x_n}{\gamma + x_{n-1}}, \quad n = 0, 1, \ldots$$

$$x_{n+1} = \frac{1 + x_n}{x_{n-1}}, \quad n = 0, 1, \ldots$$

$$x_{n+1} = \frac{\beta x_n + \gamma x_{n-1}}{A + Bx_n + Cx_{n-1}}, \quad n = 0, 1, \ldots$$

are of types $(1, 1), (2, 2), (2, 1)$, and $(2, 3)$ respectively.

In this sense, Eq(1) contains 49 equations as follows:

9 equations of type $(1, 1)$
9 equations of type $(1, 2)$
9 equations of type $(2, 1)$
9 equations of type $(2, 2)$
3 equations of type $(1, 3)$
3 equations of type $(3, 1)$
3 equations of type $(2, 3)$
3 equations of type $(3, 2)$
1 equation of type $(3, 3)$.

Of these 49 equations, 21 are either trivial, or linear, or of the Riccati type. The remaining 28 are quite interesting nonlinear difference equations. Some of them have been recently investigated and have led to the development of some general theory about difference equations. See [7], [15], [28], [29], [42], [43], [45] and [52]-[54]. However to this day, many special cases of Eq(1) are not thoroughly understood yet and remain a great challenge for further investigation.

Our goal in this monograph is to familiarize readers with the recent techniques and results related to Eq(1) and its extensions, and to bring to their attention several **Open Research Problems and Conjectures** related to this equation and its functional generalization.

Chapter 1

Preliminary Results

In this chapter we state some known results and prove some new ones about difference equations which will be useful in our investigation of Eq(1).

The reader may just glance at the results in this chapter and return for the details when they are needed in the sequel. See also [1], [4], [16], [21], [22], [33], [34], [41], [63], and [68].

1.1 Definitions of Stability and Linearized Stability Analysis

Let I be some interval of real numbers and let

$$f : I \times I \to I$$

be a continuously differentiable function.

Then for every set of **initial conditions** $x_0, x_{-1} \in I$, the difference equation

$$x_{n+1} = f(x_n, x_{n-1}), \quad n = 0, 1, \ldots \tag{1.1}$$

has a **unique solution** $\{x_n\}_{n=-1}^{\infty}$.

A point $\overline{x} \in I$ is called an **equilibrium point** of Eq(1.1) if

$$\overline{x} = f(\overline{x}, \overline{x});$$

that is,

$$x_n = \overline{x} \quad \text{for } n \geq 0$$

is a solution of Eq(1.1), or equivalently, $\overline{x}$ is a **fixed point** of f.

Definition 1.1.1 *Let $\overline{x}$ be an equilibrium point of Eq(1.1).*

(i) *The equilibrium* $\overline{x}$ *of Eq(1.1) is called* **locally stable** *if for every* $\varepsilon > 0$, *there exists* $\delta > 0$ *such that for all* $x_0, x_{-1} \in I$ *with* $|x_0 - \overline{x}| + |x_{-1} - \overline{x}| < \delta$,

we have

$$|x_n - \overline{x}| < \varepsilon \text{ for all } n \geq -1.$$

(ii) *The equilibrium* $\overline{x}$ *of Eq(1.1) is called* **locally asymptotically stable** *if it is locally stable, and if there exists* $\gamma > 0$ *such that for all* $x_0, x_{-1} \in I$ *with* $|x_0 - \overline{x}| + |x_{-1} - \overline{x}| < \gamma$,

we have

$$\lim_{n \to \infty} x_n = \overline{x}.$$

(iii) *The equilibrium* $\overline{x}$ *of Eq(1.1) is called a* **global attractor** *if for every* $x_0, x_{-1} \in I$ *we have*

$$\lim_{n \to \infty} x_n = \overline{x}.$$

(iv) *The equilibrium* $\overline{x}$ *of Eq(1.1) is called* **globally asymptotically stable** *if it is locally stable and a global attractor.*

(v) *The equilibrium* $\overline{x}$ *of Eq(1.1) is called* **unstable** *if it is not stable.*

(vi) *The equilibrium* $\overline{x}$ *of Eq(1.1) is called a* **source**, *or a* **repeller**, *if there exists* $r > 0$ *such that for all* $x_0, x_{-1} \in I$ *with* $0 < |x_0 - \overline{x}| + |x_{-1} - \overline{x}| < r$, *there exists* $N \geq 1$ *such that*

$$|x_N - \overline{x}| \geq r.$$

Clearly a source is an unstable equilibrium.

Let

$$p = \frac{\partial f}{\partial u}(\overline{x}, \overline{x}) \quad \text{and} \quad q = \frac{\partial f}{\partial v}(\overline{x}, \overline{x})$$

denote the partial derivatives of $f(u, v)$ evaluated at the equilibrium $\overline{x}$ of Eq(1.1).

Then the equation

$$y_{n+1} = py_n + qy_{n-1}, \quad n = 0, 1, \ldots \tag{1.2}$$

is called the *linearized equation associated with Eq(1.1) about the equilibrium point* $\overline{x}$.

Theorem 1.1.1 *(Linearized Stability)*

(a) *If both roots of the quadratic equation*

$$\lambda^2 - p\lambda - q = 0 \tag{1.3}$$

lie in the open unit disk $|\lambda| < 1$, *then the equilibrium* $\overline{x}$ *of Eq(1.1) is locally asymptotically stable.*

(b) *If at least one of the roots of Eq(1.3) has absolute value greater than one, then the equilibrium $\overline{x}$ of Eq(1.1) is unstable.*

(c) *A necessary and sufficient condition for both roots of Eq(1.3) to lie in the open unit disk $|\lambda| < 1$, is*

$$|p| < 1 - q < 2. \tag{1.4}$$

In this case the locally asymptotically stable equilibrium $\overline{x}$ is also called a **sink**.

(d) *A necessary and sufficient condition for both roots of Eq(1.3) to have absolute value greater than one is*

$$|q| > 1 \quad \text{and} \quad |p| < |1 - q|\,.$$

In this case $\overline{x}$ is a **repeller**.

(e) *A necessary and sufficient condition for one root of Eq(1.3) to have absolute value greater than one and for the other to have absolute value less than one is*

$$p^2 + 4q > 0 \quad \text{and} \quad |p| > |1 - q|\,.$$

In this case the unstable equilibrium $\overline{x}$ is called a **saddle point**.

(f) *A necessary and sufficient condition for a root of Eq(1.3) to have absolute value equal to one is*

$$|p| = |1 - q|$$

or

$$q = -1 \quad \text{and} \quad |p| \leq 2.$$

In this case the equilibrium $\overline{x}$ is called a **nonhyperbolic point**.

Definition 1.1.2 *(a) A solution $\{x_n\}$ of Eq(1.1) is said to be* **periodic** *with period p if*

$$x_{n+p} = x_n \quad \text{for all} \quad n \geq -1. \tag{1.5}$$

(b) A solution $\{x_n\}$ of Eq(1.1) is said to be **periodic with prime period** *p, or a p-**cycle** if it is periodic with period p and p is the least positive integer for which (1.5) holds.*

1.2 The Stable Manifold Theorem in the Plane

Let $\mathbf{T}$ be a diffeomorphism in $I \times I$, i.e., one-to-one, smooth mapping, with smooth inverse.

Assume that $\mathbf{p} \in I \times I$ is a saddle fixed point of $\mathbf{T}$, i.e., $\mathbf{T}(\mathbf{p}) = \mathbf{p}$ and the Jacobian $J_{\mathbf{T}}(\mathbf{p})$ has one eigenvalue s with $|s| < 1$ and one eigenvalue u with $|u| > 1$.

Let $\mathbf{v}_s$ be an eigenvector corresponding to s and let $\mathbf{v}_u$ be an eigenvector corresponding to u.

Let S be the **stable manifold** of $\mathbf{p}$, i.e., the set of initial points whose forward orbits (under iteration by $\mathbf{T}$)

$$\mathbf{p}, \mathbf{T}(\mathbf{p}), \mathbf{T}^2(\mathbf{p}), \ldots$$

converge to $\mathbf{p}$.

Let U be the **unstable manifold** of $\mathbf{p}$, i.e., the set of initial points whose backward orbits (under iteration by the inverse of $\mathbf{T}$)

$$\mathbf{p}, \mathbf{T}^{-1}(\mathbf{p}), \mathbf{T}^{-2}(\mathbf{p}), \ldots$$

converge to $\mathbf{p}$. Then, *both S and U are one dimensional manifolds (curves) that contain $\mathbf{p}$. Furthermore, the vectors $\mathbf{v}_s$ and $\mathbf{v}_u$ are tangent to S and U at $\mathbf{p}$, respectively.*

The map $\mathbf{T}$ for Eq(1.1) is found as follows:

Set

$$u_n = x_{n-1} \quad \text{and} \quad v_n = x_n \quad \text{for } n \geq 0.$$

Then

$$u_{n+1} = x_n \quad \text{and} \quad v_{n+1} = f(v_n, u_n) \quad \text{for } n \geq 0$$

and so

$$\mathbf{T}\begin{pmatrix} u \\ v \end{pmatrix} = \begin{pmatrix} v \\ f(v,u) \end{pmatrix}.$$

Therefore the eigenvalues of the Jacobian $J_{\mathbf{T}}\begin{pmatrix} \overline{x} \\ \overline{x} \end{pmatrix}$ at the equilibrium point $\overline{x}$ of Eq(1.1) are the roots of Eq(1.3).

In this monograph, the way we will make use of the Stable Manifold Theorem in our investigation of the rational difference equation

$$x_{n+1} = \frac{\alpha + \beta x_n + \gamma x_{n-1}}{A + Bx_n + Cx_{n-1}}, \quad n = 0, 1, \ldots \tag{1.6}$$

is as follows:

Under certain conditions the positive equilibrium of Eq(1.6) will be a saddle point and under the same conditions Eq(1.6) will have a two cycle

$$\ldots \phi, \psi, \phi, \psi, \ldots \tag{1.7}$$

which is locally asymptotically stable. Under these conditions, the two cycle (1.7) cannot be a global attractor of all non-equilibrium solutions. See, for example, Sections 6.6 and 6.9.

1.3 Global Asymptotic Stability of the Zero Equilibrium

When zero is an equilibrium point of the Eq(1.6), which we are investigating in this monograph (with $\beta > 0$), then the following theorem will be employed to establish conditions for its global asymptotic stability. Amazingly the same theorem is a powerful tool for establishing the global asymptotic stability of the zero equilibrium of several biological models. See, for example, [31] and [48].

Theorem 1.3.1 ([31]. See also [48], and [70]) *Consider the difference equation*

$$x_{n+1} = f_0(x_n, x_{n-1})x_n + f_1(x_n, x_{n-1})x_{n-1}, \quad n = 0, 1, \ldots \tag{1.8}$$

with nonnegative initial conditions and

$$f_0, f_1 \in C[[0,\infty) \times [0,\infty), [0,1)].$$

Assume that the following hypotheses hold:

(i) f_0 and f_1 are non-increasing in each of their arguments;

(ii) $f_0(x,x) > 0$ for all $x \geq 0$;

(iii) $f_0(x,y) + f_1(x,y) < 1$ for all $x, y \in (0,\infty)$.

Then the zero equilibrium of Eq(1.8) is globally asymptotically stable.

1.4 Global Attractivity of the Positive Equilibrium

Unfortunately when it comes to establishing the global attractivity of the positive equilibrium of the Eq(1.6) which we are investigating in this monograph, there are not enough results in the literature to cover all various cases. We strongly believe that the investigation of the various special cases of our equation has already played and will continue to play an important role in the development of the stability theory of general difference equations of orders greater than one.

The first theorem, which has also been very useful in applications to mathematical biology, see [31] and [48], was really motivated by a problem in [35].

Theorem 1.4.1 ([31]. See also [35], [48], and ([42], p. 53). *Let $I \subseteq [0,\infty)$ be some interval and assume that $f \in C[I \times I, (0,\infty)]$ satisfies the following conditions:*

(i) $f(x,y)$ is non-decreasing in each of its arguments;

(ii) Eq(1.1) has a unique positive equilibrium point $\overline{x} \in I$ and the function $f(x,x)$ satisfies the **negative feedback condition***:*

$$(x-\overline{x})(f(x,x)-x) < 0 \quad \textit{for every} \quad x \in I - \{\overline{x}\}.$$

Then every positive solution of Eq(1.1) with initial conditions in I converges to $\overline{x}$.

The following theorem was inspired by the results in [55] that deal with the global asymptotic stability of **Pielou's equation** (see Section 4.4), which is a special case of the equation that we are investigating in this monograph.

Theorem 1.4.2 ([42], p. 27) *Assume*
(i) $f \in C[(0,\infty)\times(0,\infty),(0,\infty)]$;
(ii) $f(x,y)$ *is nonincreasing in* x *and decreasing in* y;
(iii) $xf(x,x)$ *is increasing in* x;
(iv) The equation

$$x_{n+1} = x_n f(x_n, x_{n-1}), \quad n = 0, 1, \ldots \tag{1.9}$$

has a unique positive equilibrium $\overline{x}$.
Then $\overline{x}$ *is globally asymptotically stable.*

Theorem 1.4.3 ([27]) *Assume*
(i) $f \in C[[0,\infty)\times[0,\infty),(0,\infty)]$;
(ii) $f(x,y)$ *is nonincreasing in each argument;*
(iii) $xf(x,y)$ *is nondecreasing in* x;
(iv) $f(x,y) < f(y,x) \Longleftrightarrow x > y$;
(v) The equation

$$x_{n+1} = x_{n-1} f(x_{n-1}, x_n), \quad n = 0, 1, \ldots \tag{1.10}$$

has a unique positive equilibrium $\overline{x}$.
Then $\overline{x}$ *is a global attractor of all positive solutions of Eq(1.10).*

The next result is known as the **stability trichotomy** result:

Theorem 1.4.4 ([46]) *Assume*

$$f \in C^1[[0,\infty)\times[0,\infty),[0,\infty)]$$

is such that

$$x\left|\frac{\partial f}{\partial x}\right| + y\left|\frac{\partial f}{\partial y}\right| < f(x,y) \quad \textit{for all} \quad x, y \in (0,\infty). \tag{1.11}$$

Then Eq(1.1) has **stability trichotomy**, *that is exactly one of the following three cases holds for all solutions of Eq(1.1):*

(i) $\lim_{n\to\infty} x_n = \infty$ *for all* $(x_{-1}, x_0) \neq (0,0)$.

(ii) $\lim_{n\to\infty} x_n = 0$ *for all initial points and* 0 *is the only equilibrium point of Eq(1.1).*

(iii) $\lim_{n\to\infty} x_n = \overline{x} \in (0,\infty)$ *for all* $(x_{-1}, x_0) \neq (0,0)$ *and* $\overline{x}$ *is the only positive equilibrium of Eq(1.1).*

Very often the best strategy for obtaining global attractivity results for Eq(1.1) is to work in the regions where the function $f(x, y)$ is monotonic in its arguments. In this regard there are four possible scenarios depending on whether $f(x, y)$ is nondecreasing in both arguments, or nonincreasing in both arguments, or nonincreasing in one and nondecreasing in the other.

The next two theorems are slight modifications of results which were obtained in [54] and were developed while we were working on a special case of the equation that we are investigating in this monograph. (See Section 6.9.) The first theorem deals with the case where the function $f(x, y)$ is non-decreasing in x and non-increasing in y and the second theorem deals with the case where the function $f(x, y)$ is non-increasing in x and non-decreasing in y. Theorems 1.4.8 and 1.4.7 below are new and deal with the remaining two cases relative to the monotonic character of $f(x, y)$. See Sections 6.4 and 6.8 where these theorems are utilized. See also the Appendix, at the end of this monograph, which describes the extension of these results to higher order difference equations.

Theorem 1.4.5 ([54]) *Let $[a, b]$ be an interval of real numbers and assume that*

$$f : [a, b] \times [a, b] \to [a, b]$$

is a continuous function satisfying the following properties:

(a) $f(x, y)$ is non-decreasing in $x \in [a, b]$ for each $y \in [a, b]$, and $f(x, y)$ is non-increasing in $y \in [a, b]$ for each $x \in [a, b]$;

(b) If $(m, M) \in [a, b] \times [a, b]$ is a solution of the system

$$f(m, M) = m \quad \text{and} \quad f(M, m) = M,$$

then $m = M$.

Then Eq(1.1) has a unique equilibrium $\overline{x} \in [a, b]$ and every solution of Eq(1.1) converges to $\overline{x}$.

Proof. Set

$$m_0 = a \quad \text{and} \quad M_0 = b$$

and for $i = 1, 2, \ldots$ set

$$M_i = f(M_{i-1}, m_{i-1}) \quad \text{and} \quad m_i = f(m_{i-1}, M_{i-1}).$$

Now observe that for each $i \geq 0$,

$$m_0 \leq m_1 \leq \ldots \leq m_i \leq \ldots \leq M_i \leq \ldots \leq M_1 \leq M_0$$

and

$$m_i \leq x_k \leq M_i \quad \text{for } k \geq 2i + 1.$$

Set

$$m = \lim_{i\to\infty} m_i \quad \text{and} \quad M = \lim_{i\to\infty} M_i.$$

Then

$$M \geq \limsup_{i\to\infty}\ x_i \geq \liminf_{i\to\infty}\ x_i \geq m \tag{1.12}$$

and by the continuity of f,

$$m = f(m, M) \quad \text{and} \quad M = f(M, m).$$

Therefore in view of *(b)*,

$$m = M$$

from which the result follows. □

Theorem 1.4.6 ([54]) *Let $[a, b]$ be an interval of real numbers and assume that*

$$f : [a, b] \times [a, b] \to [a, b]$$

is a continuous function satisfying the following properties:

(a) $f(x, y)$ is non-increasing in $x \in [a, b]$ for each $y \in [a, b]$, and $f(x, y)$ is non-decreasing in $y \in [a, b]$ for each $x \in [a, b]$;

(b) The difference equation Eq(1.1) has no solutions of prime period two in $[a, b]$. Then Eq(1.1) has a unique equilibrium $\overline{x} \in [a, b]$ and every solution of Eq(1.1) converges to $\overline{x}$.

Proof. Set

$$m_0 = a \quad \text{and} \quad M_0 = b$$

and for $i = 1, 2, \ldots$ set

$$M_i = f(m_{i-1}, M_{i-1}) \quad \text{and} \quad m_i = f(M_{i-1}, m_{i-1}).$$

Now observe that for each $i \geq 0$,

$$m_0 \leq m_1 \leq \ldots \leq m_i \leq \ldots \leq M_i \leq \ldots \leq M_1 \leq M_0,$$

and

$$m_i \leq x_k \leq M_i \quad \text{for } k \geq 2i + 1.$$

Set

$$m = \lim_{i\to\infty} m_i \quad \text{and} \quad M = \lim_{i\to\infty} M_i.$$

Then clearly (1.12) holds and by the continuity of f,

$$m = f(M, m) \quad \text{and} \quad M = f(m, M).$$

In view of *(b)*,

$$m = M$$

from which the result follows. □

Theorem 1.4.7 *Let* $[a,b]$ *be an interval of real numbers and assume that*

$$f : [a,b] \times [a,b] \to [a,b]$$

is a continuous function satisfying the following properties:

(a) $f(x,y)$ *is non-increasing in each of its arguments;*

(b) If $(m,M) \in [a,b] \times [a,b]$ *is a solution of the system*

$$f(m,m) = M \quad \text{and} \quad f(M,M) = m,$$

then $m = M$.

Then Eq(1.1) has a unique equilibrium $\overline{x} \in [a,b]$ *and every solution of Eq(1.1) converges to* $\overline{x}$.

Proof. Set

$$m_0 = a \quad \text{and} \quad M_0 = b$$

and for $i = 1, 2, \ldots$ set

$$M_i = f(m_{i-1}, m_{i-1}) \quad \text{and} \quad m_i = f(M_{i-1}, M_{i-1}).$$

Now observe that for each $i \geq 0$,

$$m_0 \leq m_1 \leq \ldots \leq m_i \leq \ldots \leq M_i \leq \ldots \leq M_1 \leq M_0,$$

and

$$m_i \leq x_k \leq M_i \quad \text{for } k \geq 2i+1.$$

Set

$$m = \lim_{i\to\infty} m_i \quad \text{and} \quad M = \lim_{i\to\infty} M_i.$$

Then clearly (1.12) holds and by the continuity of f,

$$m = f(M,M) \quad \text{and} \quad M = f(m,m).$$

In view of *(b)*,

$$m = M = \overline{x}$$

from which the result follows. □

Theorem 1.4.8 *Let* $[a,b]$ *be an interval of real numbers and assume that*

$$f : [a,b] \times [a,b] \to [a,b]$$

is a continuous function satisfying the following properties:

(a) $f(x,y)$ *is non-decreasing in each of its arguments;*

(b) The equation

$$f(x,x) = x,$$

has a unique positive solution.

Then Eq(1.1) has a unique equilibrium $\overline{x} \in [a,b]$ *and every solution of Eq(1.1) converges to* $\overline{x}$.

Proof. Set

$$m_0 = a \quad \text{and} \quad M_0 = b$$

and for $i = 1, 2, \ldots$ set

$$M_i = f(M_{i-1}, M_{i-1}) \quad \text{and} \quad m_i = f(m_{i-1}, m_{i-1}).$$

Now observe that for each $i \geq 0$,

$$m_0 \leq m_1 \leq \ldots \leq m_i \leq \ldots \leq M_i \leq \ldots \leq M_1 \leq M_0,$$

and

$$m_i \leq x_k \leq M_i \quad \text{for } k \geq 2i + 1.$$

Set

$$m = \lim_{i \to \infty} m_i \quad \text{and} \quad M = \lim_{i \to \infty} M_i.$$

Then clearly the inequality (1.12) is satisfied and by the continuity of f,

$$m = f(m, m) \quad \text{and} \quad M = f(M, M).$$

In view of *(b)*,

$$m = M = \overline{x},$$

from which the result follows. □

1.5 Limiting Solutions

The concept of limiting solutions was introduced by Karakostas [37] (see also [40]) and is a useful tool in establishing that under certain conditions a solution of a difference equation which is bounded from above and from below has a limit.

In this section we first give a self-contained version of limiting solutions the way we will use them in this monograph.

Let J be some interval of real numbers, $f \in C[J \times J, J]$, and let $k \geq 1$ be a positive integer. Assume that $\{x_n\}_{n=-1}^{\infty}$ is a solution of Eq(1.1) such that

$$A \leq x_n \leq B \quad \text{for all } n \geq -1,$$

where $A, B \in J$. Let L be a limit point of the solution $\{x_n\}_{n=-1}^{\infty}$ (For example L is the limit superior S or the limit inferior I of the solution). Now with the above hypotheses, we will show that there exists a solution $\{L_n\}_{n=-k}^{\infty}$ of the difference equation

$$x_{n+1} = f(x_n, x_{n-1}), \quad n \geq -k + 1, \ldots \tag{1.13}$$

such that

$$L_0 = L$$

and for every $N \geq -k$, the term L_N is a limit point of the solution $\{x_n\}_{n=-1}^{\infty}$.

The solution $\{L_n\}_{n=-k}^{\infty}$ is called a **limiting solution** of Eq(1.13) associated with the solution $\{x_n\}_{n=-1}^{\infty}$ of Eq(1.1).

Karakostas takes $k = \infty$ and calls the solution $\{L_n\}_{n=-\infty}^{\infty}$ a **full limiting sequence** of the difference equation

$$x_{n+1} = f(x_n, x_{n-1}), \tag{1.14}$$

associated with the solution $\{x_n\}_{n=-1}^{\infty}$ of Eq(1.1). (See Theorem 1.5.2 below.)

We now proceed to show the existence of a limiting solution $\{L_n\}_{n=-k}^{\infty}$.

Clearly there exists a subsequence $\{x_{n_i}\}_{i=1}^{\infty}$ of $\{x_n\}_{n=-1}^{\infty}$ such that

$$\lim_{i\to\infty} x_{n_i} = L.$$

Next the subsequence $\{x_{n_i-1}\}_{i=1}^{\infty}$ of $\{x_n\}_{n=-1}^{\infty}$ also lies in interval $[A, B]$ and so it has a further subsequence $\{x_{n_{i_j}}\}_{j=1}^{\infty}$ which converges to some limit which we call L_{-1}. Therefore,

$$\lim_{i\to\infty} x_{n_{i_j}} = L \quad \text{and} \quad \lim_{i\to\infty} x_{n_{i_j}-1} = L_{-1}.$$

Continuing in this way and by considering further and further subsequences we obtain numbers $L_{-2}, L_{-3}, \ldots, L_{-k}$ and a subsequence $\{x_{n_j}\}_{n=-1}^{\infty}$ such that

$$\lim_{j\to\infty} x_{n_j} = L$$

$$\lim_{j\to\infty} x_{n_j-1} = L_{-1}$$

$$\ldots$$

$$\lim_{j\to\infty} x_{n_j-k} = L_{-k}.$$

Let $\{\ell_n\}_{n=-k}^{\infty}$ be the solution of Eq(1.13) with $\ell_{-k} = L_{-k}$ and $\ell_{-k+1} = L_{-k+1}$.

Then clearly,

$$\ell_{-k+2} = f(\ell_{-k+1}, \ell_{-k}) = f(L_{-k+1}, L_{-k})$$
$$= \lim_{j\to\infty} f(x_{n_j-k+1}, x_{n_j-k}) = \lim_{j\to\infty} x_{n_j-k+2} = L_{-k+2}.$$

It follows by induction that the sequence

$$L_{-k}, L_{-k+1}, \ldots, L_{-1}, L_0 = L = \ell_0, L_1 = \ell_1, \ldots$$

satisfies Eq(1.13) and every term of this sequence is a limit point of the solution $\{x_n\}_{n=-1}^{\infty}$.

The following result, which is a consequence of the above discussion, is stated in the way we will use it in this monograph:

Theorem 1.5.1 *Let J be some interval of real numbers, $f \in C[J \times J, J]$, and let $k \geq 1$ be a positive integer. Suppose that $\{x_n\}_{n=-1}^{\infty}$ is a bounded solution of Eq(1.1) with limit inferior I and limit superior S, with $I, S \in J$. Then Eq(1.13) has two solutions $\{I_n\}_{n=-k}^{\infty}$ and $\{S_n\}_{n=-k}^{\infty}$ with the following properties:*

$$I_0 = I, \quad S_0 = S$$

and

$$I_n, S_n \in [I, S] \quad \text{for } n \geq -k.$$

We now state the complete version of full limiting sequences, as developed by Karakostas.

Theorem 1.5.2 ([37]) *Let J be some interval of real numbers, $f \in C[J^{\nu+1}, J]$, and let $\{x_n\}_{n=-\nu}^{\infty}$ be a bounded solution of the difference equation*

$$x_{n+1} = f(x_n, x_{n-1}, \ldots, x_{n-\nu}), \quad n = 0, 1, \ldots \tag{1.15}$$

with

$$I = \liminf_{n\to\infty} x_n, \quad S = \limsup_{n\to\infty} x_n \quad \text{and with} \quad I, S \in J.$$

Let Z denote the set of all integers $\{\ldots, -1, 0, 1, \ldots\}$. Then there exist two solutions $\{I_n\}_{n=-\infty}^{\infty}$ and $\{S_n\}_{n=-\infty}^{\infty}$ of the difference equation

$$x_{n+1} = f(x_n, x_{n-1}, \ldots, x_{n-\nu}) \tag{1.16}$$

which satisfy the equation for all $n \in Z$, with

$$I_0 = I, S_0 = S, I_n, S_n \in [I, S] \quad \text{for all } n \in Z$$

and such that for every $N \in Z$, I_N and S_N are limit points of $\{x_n\}_{n=-\nu}^{\infty}$.

Furthermore for every $m \leq -\nu$, there exist two subsequences $\{x_{r_n}\}$ and $\{x_{l_n}\}$ of the solution $\{x_n\}_{n=-\nu}^{\infty}$ such that the following are true:

$$\lim_{n\to\infty} x_{r_n+N} = I_N \quad \text{and} \quad \lim_{n\to\infty} x_{l_n+N} = S_N \quad \text{for every } N \geq m.$$

The solutions $\{I_n\}_{n=-\infty}^{\infty}$ and $\{S_n\}_{n=-\infty}^{\infty}$ of the difference equation (1.16) are called **full limiting solutions** of Eq(1.16) associated with the solution $\{x_n\}_{n=-k}^{\infty}$ of Eq(1.15). See [23] and [25] for an application of Theorem 1.5.2 to difference equations of the form

$$x_{n+1} = \sum_{i=0}^{k} \frac{A_i}{x_{n-i}}, \quad n = 0, 1, \ldots$$

which by the change of variables $x_n = \frac{1}{y_n}$ reduce to rational difference equations of the form

$$y_{n+1} = \frac{1}{\sum_{i=0}^{k} A_i y_{n-i}}, \quad n = 0, 1, \ldots .$$

1.6 The Riccati Equation

A difference equation of the form

$$x_{n+1} = \frac{\alpha + \beta x_n}{A + Bx_n}, \quad n = 0, 1, \ldots \tag{1.17}$$

where the parameters α, β, A, B and the initial condition x_0 are real numbers is called a **Riccati difference equation**.

To avoid degenerate cases, we will assume that

$$B \neq 0 \quad \text{and} \quad \alpha B - \beta A \neq 0. \tag{1.18}$$

Indeed when $B = 0$, Eq(1.17) is a linear equation and when $\alpha B - \beta A = 0$, Eq(1.17) reduces to

$$x_{n+1} = \frac{\frac{\beta A}{B} + \beta x_n}{A + Bx_n} = \frac{\beta}{B} \quad \text{for all} \quad n \geq 0.$$

The set of initial conditions $x_0 \in \mathbf{R}$ through which the denominator $A + Bx_n$ in Eq(1.17) will become zero for some value of $n \geq 0$ is called the **forbidden set F** of Eq(1.17).

One of our goals in this section is to determine the forbidden set $\mathbf{F}$ of Eq(1.17). It should be mentioned here that there is practically nothing known about the forbidden set of Eq(1) (when the parameters $\alpha, \beta, \gamma, A, B, C$ and the initial conditions x_{-1}, x_0 are all real numbers) other than the material presented here, which is extracted from [32]. (See also [13] and [69].)

The other goal we have is to describe in detail the long- and short-term behavior of solutions of Eq(1.17) when $x_0 \notin \mathbf{F}$.

The first result in this section addresses a very special case of Eq(1.17) regarding period-two solutions. The proof is straightforward and will be omitted.

Theorem 1.6.1 *Assume that (1.18) holds and that Eq(1.17) possesses a prime period-two solution. Then*

$$\beta + A = 0. \tag{1.19}$$

Furthermore when (1.19) holds every solution of Eq(1.17) with

$$x_0 \neq \frac{\beta}{B} \tag{1.20}$$

is periodic with period two.

In the sequel, in addition to (1.18), we will assume that

$$\beta + A \neq 0. \tag{1.21}$$

Now observe that the change of variable

$$x_n = \frac{\beta + A}{B} w_n - \frac{A}{B} \quad \text{for} \quad n \geq 0 \tag{1.22}$$

transforms Eq(1.17) into the difference equation

$$w_{n+1} = 1 - \frac{R}{w_n}, \quad n = 0, 1, \ldots \tag{1.23}$$

where

$$R = \frac{\beta A - \alpha B}{(\beta + A)^2}$$

is a nonzero real number, called the **Riccati number** of Eq(1.17).

It is interesting to note that the four parameters of Eq(1.17) have been reduced to the single parameter R of Eq(1.23). If we set

$$y = g(w) = 1 - \frac{R}{w}$$

then we note that the two asymptotes

$$x = -\frac{A}{B} \quad \text{and} \quad y = \frac{\beta}{B}$$

of the function

$$y = f(x) = \frac{\alpha + \beta x}{A + Bx}$$

have now been reduced to the two asymptotes

$$w = 0 \quad \text{and} \quad y = 1$$

of the function

$$y = 1 - \frac{R}{w}.$$

Also note that the signum of R is equal to the signum of the derivative

$$f'(x) = \frac{\beta A - \alpha B}{(A + Bx)^2}.$$

For the remainder of this section we focus on Eq(1.23) and present its forbidden set **F** and the character of its solutions. Similar results can be obtained for Eq(1.17) through the change of variables (1.22).

From Eq(1.23) it follows that when

$$R < \frac{1}{4}$$

Eq(1.23) has the two equilibrium points w_- and w_+ with

$$w_- = \frac{1-\sqrt{1-4R}}{2} < \frac{1+\sqrt{1-4R}}{2} = w_+$$

and

$$|w_-| < w_+.$$

Also

$$w_- < 0 \quad \text{if} \quad R < 0$$

and

$$w_- > 0 \quad \text{if} \quad R \in (0, \frac{1}{4}).$$

When

$$R = \frac{1}{4},$$

Eq(1.23) has exactly one equilibrium point namely,

$$\overline{w} = \frac{1}{2}.$$

Finally, when

$$R > \frac{1}{4},$$

Eq(1.23) has no equilibrium points.

Now observe that the change of variables

$$w_n = \frac{y_{n+1}}{y_n} \quad \text{for} \quad n = 0, 1, \ldots$$

with

$$y_0 = 1 \quad \text{and} \quad y_1 = w_0$$

reduces Eq(1.23) to the second order linear difference equation

$$y_{n+2} - y_{n+1} + Ry_n = 0, \quad n = 0, 1, \ldots \tag{1.24}$$

which can be solved explicitly.

The forbidden set of Eq(1.23) is clearly the set of points where the sequence $\{y_n\}_{n=0}^{\infty}$ vanishes.

The next two theorems deal with the cases $R < \frac{1}{4}$ and $R = \frac{1}{4}$, respectively, where the characteristic roots of Eq(1.24) are real numbers. Note that in these two cases the characteristic roots of Eq(1.24) are the equilibrium points of Eq(1.23).

Theorem 1.6.2 *Assume*

$$R < \frac{1}{4}.$$

Then

$$w_- = \frac{1 - \sqrt{1-4R}}{2} \quad \textit{and} \quad w_+ = \frac{1 + \sqrt{1-4R}}{2}$$

are the only equilibrium points of Eq(1.23).

The forbidden set **F** *of Eq(1.23) is the sequence of points*

$$f_n = \left(\frac{w_+^{n-1} - w_-^{n-1}}{w_+^n - w_-^n}\right) w_+ w_- \quad \textit{for} \quad n = 1, 2, \dots . \tag{1.25}$$

When $w_0 \notin \mathbf{F}$, the solution of Eq(1.23) is given by

$$w_n = \frac{(w_0 - w_-)w_+^{n+1} - (w_+ - w_0)w_-^{n+1}}{(w_0 - w_-)w_+^n - (w_+ - w_0)w_-^n}, \quad n = 1, 2, \dots . \tag{1.26}$$

It is now clear from Theorem 1.6.2 that

$$\lim_{n\to\infty} f_n = w_-,$$

and

$$f_n < w_- \quad \text{if} \quad R > 0$$

while

$$(f_n - w_-)(f_{n+1} - w_-) < 0 \quad \text{if} \quad R < 0.$$

The equilibrium w_+ is a global attractor as long as

$$w_0 \notin F \cup \{w_-\}$$

and the equilibrium w_- is a repeller.

Theorem 1.6.3 *Assume*

$$R = \frac{1}{4}.$$

Then

$$\overline{w} = \frac{1}{2}$$

is the only equilibrium point of Eq(1.23).

The forbidden set **F** *of Eq(1.23) is the sequence of points*

$$f_n = \frac{n-1}{2n} \quad \textit{for} \quad n = 1, 2, \dots , \tag{1.27}$$

which converges to the equilibrium from the left.
When $w_0 \notin \mathbf{F}$, the solution of Eq(1.23) is given by

$$w_n = \frac{1 + (2w_0 - 1)(n+1)}{2 + 2(2w_0 - 1)n} \quad \textit{for} \quad n = 0, 1, \ldots . \tag{1.28}$$

From Theorem 1.6.3 it follows that the equilibrium $\bar{w} = \frac{1}{2}$ of Eq(1.23) is a global attractor for all solutions with $w_0 \notin \mathbf{F}$ but is locally unstable, more precisely, it is a sink from the right and a source from the left.

Finally consider the case

$$R > \frac{1}{4}.$$

The characteristic roots of Eq(1.24) in this case are

$$\lambda_{\pm} = \frac{1}{2} \pm i\frac{\sqrt{4R-1}}{2}$$

with

$$|\lambda_{\pm}| = \sqrt{R}.$$

Let

$$\phi \in (0, \frac{\pi}{2})$$

be such that

$$\cos\phi = \frac{1}{2\sqrt{R}} \quad \text{and} \quad \sin\phi = \frac{\sqrt{4R-1}}{2\sqrt{R}}.$$

Then

$$y_n = R^{\frac{n}{2}}(C_1 \cos(n\phi) + C_2 \sin(n\phi)), \quad n = 0, 1, \ldots$$

with

$$C_1 = y_0 = 1$$

and

$$\sqrt{R}(C_1 \cos\phi + C_2 \sin\phi) = y_1 = w_0.$$

It follows that

$$C_1 = 1 \quad \text{and} \quad C_2 = \frac{2w_0 - 1}{\sqrt{4R-1}}.$$

Hence

$$w_n = \sqrt{R}\frac{\sqrt{4R-1}\cos(n+1)\phi + (2w_0 - 1)\sin(n+1)\phi}{\sqrt{4R-1}\cos(n\phi) + (2w_0 - 1)\sin(n\phi)}, \quad n = 0, 1, \ldots \tag{1.29}$$

as long as the denominator is different from zero, that is,

$$w_0 \neq \frac{1}{2} - \frac{\sqrt{4R-1}}{2}\cot(n\phi) \quad \text{for} \quad n = 1, 2, \ldots .$$

The following theorem summarizes the above discussion.

Theorem 1.6.4 *Assume that*

$$R > \frac{1}{4}$$

and let $\phi \in (0, \frac{\pi}{2})$ *be such that*

$$\cos\phi = \frac{1}{2\sqrt{R}} \quad \textit{and} \quad \sin\phi = \frac{\sqrt{4R-1}}{2\sqrt{R}}.$$

Then the forbidden set $\mathbf{F}$ *of Eq(1.23) is the sequence of points*

$$f_n = \frac{1}{2} - \frac{\sqrt{4R-1}}{2}\cot(n\phi) \quad \textit{for} \quad n = 1, 2, \ldots . \tag{1.30}$$

When $w_0 \notin \mathbf{F}$, *the solution of Eq(1.23) is given by (1.29).*

Clearly when $R > \frac{1}{4}$ the character of the solution $\{w_n\}$ and the forbidden set $\mathbf{F}$ depends on whether or not the number $\phi \in (0, \frac{\pi}{2})$ is a rational multiple of π.

Let $\theta \in (-\frac{\pi}{2}, \frac{\pi}{2})$ be such that

$$\cos\theta = \frac{\sqrt{4R-1}}{r} \quad \text{and} \quad \sin\theta = \frac{2x_0 - 1}{r}$$

where

$$r = \sqrt{(4R-1)^2 + (2x_0 - 1)^2}.$$

Then we obtain,

$$w_n = \sqrt{R}\frac{\cos(n\phi + \phi - \theta)}{\cos(n\phi - \theta)} = \sqrt{R}\frac{\cos(n\phi - \theta)\cos\phi - \sin(n\phi - \theta)\sin\phi}{cos(n\phi - \theta)}$$

$$= \sqrt{R}(\cos\phi - \sin\phi\tan(n\phi - \theta))$$

and so

$$w_n = \frac{1}{2} - \frac{\sqrt{4R-1}}{2}\tan(n\phi - \theta), \quad n = 0, 1, \ldots \tag{1.31}$$

from which the character of the solutions of Eq(1.23), when $R > \frac{1}{4}$, can now be easily revealed.

For example, assume ϕ is a rational multiple of π, that is,

$$\phi = \frac{k}{N}\pi \in (0, \frac{\pi}{2}) \tag{1.32}$$

for some

$$k, N \in \{1, 2, \ldots\} \quad \text{with } 2k < N.$$

Then

$$w_N = \frac{1}{2} - \frac{\sqrt{4R-1}}{2}\tan(k\pi - \theta) = \frac{1}{2} - \frac{\sqrt{4R-1}}{2}\tan(-\theta) = w_0$$

and so every solution of Eq(1.23) is periodic with period N.

Equivalently assume that

$$A = \frac{1}{4\cos^2\phi} \quad \text{with} \quad \phi = \frac{k}{N}\pi$$

where

$$k, N \in \{1, 2, \ldots\} \quad \text{and} \quad 2k < N,$$

then every solution of Eq(1.23) is periodic with period N.

The following theorem summarizes the above discussion.

Theorem 1.6.5 *Assume that*

$$\phi = \frac{k}{p}\pi \in (0, \frac{\pi}{2})$$

where k and p are positive comprimes and suppose that

$$\cos\phi = \frac{1}{2\sqrt{R}} \quad \textit{and} \quad \sin\phi = \frac{\sqrt{4R-1}}{2\sqrt{R}},$$

or equivalently

$$4R\cos^2(\frac{k}{p}\pi) = 1.$$

Then every solution of Eq(1.23) with

$$w_0 \neq \frac{\sin(n\frac{k}{p}\pi) - \sqrt{4R-1}\cos(n\frac{k}{p}\pi)}{2\sin(n\frac{k}{p}\pi)} \quad \textit{for} \quad n = 1, 2, \ldots, p-1$$

is periodic with period p.

When the number ϕ in Theorem 1.6.5 is not a rational multiple of π, then the following result is true:

Theorem 1.6.6 *Assume that the number ϕ in Theorem 1.6.5 is not a rational multiple of π. Then the following statements are true:*

(i) No solution of Eq(1.23) is periodic.

(ii) The set of limit points of a solution of Eq(1.23) with $w_0 \notin \mathbf{F}$ is dense in R.

Proof. Statement *(i)* is true because,

$$w_k = w_l$$

if and only if

$$\tan(k\phi - \theta) = \tan(l\phi - \theta)$$

if and only if for some integer N

$$k\phi - \theta - (l\phi - \theta) = N\pi$$

if and only if

$$(k - l)\phi = N\pi$$

that is,

$$\phi \text{ is a rational multiple of } \pi.$$

The proof of statement *(ii)* is a consequence of the form of the solution of Eq(1.23) and the fact that when ϕ is not a rational multiple of π, then the values

$$(n\phi - \theta)(\text{mod } \pi) \quad \text{for } n = 0, 1, \ldots$$

are dense in the interval $(-\frac{\pi}{2}, \frac{\pi}{2})$. □

1.7 Semicycle Analysis

We strongly believe that a semicycle analysis of the solutions of a scalar difference equation is a powerful tool for a detailed understanding of the entire character of solutions and often leads to straightforward proofs of their long term behavior.

In this section, we present some results about the semicycle character of solutions of some general difference equations of the form

$$x_{n+1} = f(x_n, x_{n-1}), \quad n = 0, 1, \ldots \tag{1.33}$$

under appropriate hypotheses on the function f.

First we give the definitions for the positive and negative semicycle of a solution of Eq(1.33) relative to an equilibrium point $\bar{x}$.

A **positive semicycle** of a solution $\{x_n\}$ of Eq(1.1) consists of a "string" of terms $\{x_l, x_{l+1}, \ldots, x_m\}$, all greater than or equal to the equilibrium $\bar{x}$, with $l \geq -1$ and $m \leq \infty$ and such that

$$\text{either } l = -1, \quad \text{or } l > -1 \text{ and } x_{l-1} < \bar{x}$$

and

$$\text{either } m = \infty, \quad \text{or } m < \infty \text{ and } x_{m+1} < \bar{x}.$$

A **negative semicycle** of a solution $\{x_n\}$ of Eq(1.33) consists of a "string" of terms $\{x_l, x_{l+1}, \ldots, x_m\}$, all less than the equilibrium $\overline{x}$, with $l \geq -1$ and $m \leq \infty$ and such that

$$\text{either } l = -1, \quad \text{or } l > -1 \quad \text{and } x_{l-1} \geq \overline{x}$$

and

$$\text{either } m = \infty, \quad \text{or} \quad m < \infty \text{ and } x_{m+1} \geq \overline{x}.$$

Definition 1.7.1 *(Oscillation)*

(a) A sequence $\{x_n\}$ is said to **oscillate about zero** *or simply to* **oscillate** *if the terms x_n are neither eventually all positive nor eventually all negative. Otherwise the sequence is called* **nonoscillatory**. *A sequence $\{x_n\}$ is called* **strictly oscillatory** *if for every $n_0 \geq 0$, there exist $n_1, n_2 \geq n_0$ such that $x_{n_1}x_{n_2} < 0$.*

(b) A sequence $\{x_n\}$ is said to **oscillate about $\overline{\mathbf{x}}$** *if the sequence $x_n - \overline{x}$ oscillates. The sequence $\{x_n\}$ is called* **strictly oscillatory about $\overline{\mathbf{x}}$** *if the sequence $x_n - \overline{x}$ is strictly oscillatory.*

The first result was established in [28] while we were working on a special case of the equation we are investigating in this monograph. (See Section 6.5.)

Theorem 1.7.1 ([28]) *Assume that $f \in C[(0,\infty) \times (0,\infty), (0,\infty)]$ is such that:*

$f(x,y)$ is decreasing in x for each fixed y, and $f(x,y)$ is increasing in y for each fixed x.

Let $\overline{x}$ be a positive equilibrium of Eq(1.33). Then except possibly for the first semicycle, every solution of Eq(1.33) has semicycles of length one.

Proof. Let $\{x_n\}$ be a solution of Eq(1.33) with at least two semicycles. Then there exists $N \geq 0$ such that either

$$x_{N-1} < \overline{x} \leq x_N$$

or

$$x_{N-1} \geq \overline{x} > x_N.$$

We will assume that

$$x_{N-1} < \overline{x} \leq x_N.$$

All other cases are similar and will be omitted. Then by using the monotonic character of $f(x,y)$ we have

$$x_{N+1} = f(x_N, x_{N-1}) < f(\overline{x}, \overline{x}) = \overline{x}$$

and

$$x_{N+2} = f(x_{N+1}, x_N) > f(\overline{x}, \overline{x}) = \overline{x}.$$

Thus

$$x_{N+1} < \overline{x} < x_{N+2}$$

and the proof follows by induction. □

The next result applies when the function f is decreasing in both arguments.

Theorem 1.7.2 *Assume that $f \in C[(0,\infty)\times(0,\infty),(0,\infty)]$ and that $f(x,y)$ is decreasing in both arguments.*

Let $\overline{x}$ be a positive equilibrium of Eq(1.33). Then every oscillatory solution of Eq(1.33) has semicycles of length at most two.

Proof. Assume that $\{x_n\}$ is an oscillatory solution with two consecutive terms x_{N-1} and x_N in a positive semicycle

$$x_{N-1} \geq \overline{x} \quad \text{and} \quad x_N \geq \overline{x}$$

with at least one of the inequalities being strict. The proof in the case of negative semicycle is similar and is omitted. Then by using the decreasing character of f we obtain:

$$x_{N+1} = f(x_N, x_{N-1}) < f(\overline{x}, \overline{x}) = \overline{x}$$

which completes the proof. □

The next result applies when the function f is increasing in both arguments.

Theorem 1.7.3 *Assume that $f \in C[(0,\infty)\times(0,\infty),(0,\infty)]$ and that $f(x,y)$ is increasing in both arguments.*

Let $\overline{x}$ be a positive equilibrium of Eq(1.33). Then except possibly for the first semicycle, every oscillatory solution of Eq(1.33) has semicycles of length one.

Proof. Assume that $\{x_n\}$ is an oscillatory solution with two consecutive terms x_{N-1} and x_N in a positive semicycle

$$x_{N-1} \geq \overline{x} \quad \text{and} \quad x_N \geq \overline{x},$$

with at least one of the inequalities being strict. The proof in the case of negative semicycle is similar and is omitted. Then by using the increasing character of f we obtain:

$$x_{N+1} = f(x_N, x_{N-1}) > f(\overline{x}, \overline{x}) = \overline{x}$$

which shows that the next term x_{N+1} also belongs to the positive semicycle. It follows by induction that all future terms of this solution belong to this positive semicycle, which is a contradiction. □

The next result applies when the function $f(x,y)$ is increasing in x and decreasing in y.

Theorem 1.7.4 *Assume that $f \in C[(0,\infty) \times (0,\infty), (0,\infty)]$ is such that:*

$f(x,y)$ is increasing in x for each fixed y, and $f(x,y)$ is decreasing in y for each fixed x.

Let $\overline{x}$ be a positive equilibrium of Eq(1.33).

Then, except possibly for the first semicycle, every oscillatory solution of Eq(1.33) has semicycles of length at least two.

Furthermore, if we assume that

$$f(u,u) = \overline{x} \quad \text{for every } u \tag{1.34}$$

and

$$f(x,y) < x \quad \text{for every } x > y > \overline{x} \tag{1.35}$$

then $\{x_n\}$ oscillates about the equilibrium $\overline{x}$ with semicycles of length two or three, except possibly for the first semicycle which may have length one. The extreme in each semicycle occurs in the first term if the semicycle has two terms, and in the second term if the semicycle has three terms.

Proof. Assume that $\{x_n\}$ is an oscillatory solution with two consecutive terms x_{N-1} and x_N such that

$$x_{N-1} < \overline{x} \leq x_N.$$

Then by using the increasing character of f we obtain

$$x_{N+1} = f(x_N, x_{N-1}) > f(\overline{x}, \overline{x}) = \overline{x}$$

which shows that the next term x_{N+1} also belongs to the positive semicycle. The proof in the case

$$x_{N-1} \geq \overline{x} > x_N,$$

is similar and is omitted.

Now also assume that $\{x_n\}$ is an oscillatory solution with two consecutive terms x_{N-1} and x_N such that

$$x_{N-1} > x_N \geq \overline{x} \quad \text{and} \quad x_{N-2} < \overline{x}.$$

Then by using the increasing character of f and Condition (1.34) we obtain

$$x_{N+1} = f(x_N, x_{N-1}) < f(x_N, x_N) = \overline{x}$$

which shows that the positive semicycle has length two. If

$$x_N > x_{N-1} > \overline{x}$$

then by using the increasing character of f and Condition (1.34) we obtain:

$$x_{N+1} = f(x_N, x_{N-1}) > f(x_N, x_N) = \overline{x}$$

and by using Condition (1.35) we find

$$x_{N+1} = f(x_N, x_{N-1}) < x_N$$

and the proof is complete.

□

Condition (1.34) is not enough to guarantee the above result. If, instead of Condition (1.35), we assume

$$f(x,y) > x \quad \text{for every } x > y > \overline{x}. \tag{1.36}$$

then Eq(1.33) has monotonic solutions.

Theorem 1.7.5 *Assume that $f \in C[(0,\infty) \times (0,\infty), (0,\infty)]$ is such that:*

$f(x,y)$ is increasing in x for each fixed y, and $f(x,y)$ is decreasing in y for each fixed x.

Let $\overline{x}$ be a positive equilibrium of Eq(1.33). Assume that f satisfies Conditions (1.34) and (1.36).

Then if $x_0 > x_{-1} > \overline{x}$, the corresponding solution is increasing.

If, on the other hand, f satisfies Condition (1.34), and

$$f(x,y) < x \quad \text{for every } x < y < \overline{x} \tag{1.37}$$

then if $x_0 < x_{-1} < \overline{x}$, the corresponding solution is decreasing.

Proof. First, assume that Condition (1.36) is satisfied and that $x_0 > x_{-1} > \overline{x}$. Then by the monotonic character of f and (1.34), we obtain

$$x_1 = f(x_0, x_{-1}) > f(x_{-1}, x_{-1}) = \overline{x}.$$

In view of Condition (1.36) we get

$$x_1 = f(x_0, x_{-1}) > x_0.$$

and the proof follows by induction.

The proof of the remaining case is similar and will be omitted. □

Chapter 2

Local Stability, Semicycles, Periodicity, and Invariant Intervals

2.1 Equilibrium Points

Here we investigate the equilibrium points of the general equation

$$x_{n+1} = \frac{\alpha + \beta x_n + \gamma x_{n-1}}{A + Bx_n + Cx_{n-1}}, \quad n = 0, 1, \dots \tag{2.1}$$

where

$$\alpha, \beta, \gamma, A, B, C \in [0, \infty) \quad \text{with} \quad \alpha + \beta + \gamma, B + C \in (0, \infty) \tag{2.2}$$

and where the initial conditions x_{-1} and x_0 are arbitrary nonnegative real numbers such that the right hand side of Eq(2.1) is well defined for all $n \geq 0$.

In view of the above restriction on the initial conditions of Eq(2.1), the equilibrium points of Eq(2.1) are the *nonnegative* solutions of the equation

$$\overline{x} = \frac{\alpha + (\beta + \gamma)\overline{x}}{A + (B + C)\overline{x}} \tag{2.3}$$

or equivalently

$$(B + C)\overline{x}^2 - (\beta + \gamma - A)\overline{x} - \alpha = 0. \tag{2.4}$$

Zero is an equilibrium point of Eq(2.1) if and only if

$$\alpha = 0 \quad \text{and} \quad A > 0. \tag{2.5}$$

When (2.5) holds, in addition to the zero equilibrium , Eq(2.1) has a positive equilibrium if and only if

$$\beta + \gamma > A.$$

In fact in this case the positive equilibrium of Eq(2.1) is unique and is given by

$$\overline{x} = \frac{\beta + \gamma - A}{B + C}. \tag{2.6}$$

When

$$\alpha = 0 \quad \text{and} \quad A = 0$$

the only equilibrium point of Eq(2.1) is positive and is given by

$$\overline{x} = \frac{\beta + \gamma}{B + C}. \tag{2.7}$$

Note that in view of Condition (2.2), when $\alpha = 0$, the quantity $\beta + \gamma$ is positive.

Finally when

$$\alpha > 0$$

the only equilibrium point of Eq(2.1) is the positive solution

$$\overline{x} = \frac{\beta + \gamma - A + \sqrt{(\beta + \gamma - A)^2 + 4\alpha(B + C)}}{2(B + C)} \tag{2.8}$$

of the quadratic equation (2.4).

If we had allowed negative initial conditions then the negative solution of Eq(2.4) would also be an equilibrium worth investigating. The problem of investigating Eq(2.1) when the initial conditions and the coefficients of the equation are real numbers is of great mathematical importance and, except for our presentation in Section 1.6 about the Riccati equation

$$x_{n+1} = \frac{\alpha + \beta x_n}{A + Bx_n}, \quad n = 0, 1, \ldots \quad ,$$

there is very little known. (See [13], [18], [32], and [69] for some results in this regard.)

In summary, it is interesting to observe that when Eq(2.1) has a positive equilibrium $\overline{x}$, then $\overline{x}$ is unique, it satisfies Eqs(2.3) and (2.4), and it is given by (2.8). This observation simplifies the investigation of the local stability of the positive equilibrium of Eq(2.1).

2.2 Stability of the Zero Equilibrium

Here we investigate the stability of the zero equilibrium of Eq(2.1).

Set

$$f(u, v) = \frac{\alpha + \beta u + \gamma v}{A + Bu + Cv}$$

and observe that

$$f_u(u, v) = \frac{(\beta A - \alpha B) + (\beta C - \gamma B)v}{(A + Bu + Cv)^2} \tag{2.9}$$

and

$$f_v(u,v) = \frac{(\gamma A - \alpha C) - (\beta C - \gamma B)u}{(A + Bu + Cv)^2}. \tag{2.10}$$

If $\overline{x}$ denotes an equilibrium point of Eq(2.1), then the linearized equation associated with Eq(2.1) about the equilibrium point $\overline{x}$ is

$$z_{n+1} - pz_n - qz_{n-1} = 0$$

where

$$p = f_u(\overline{x}, \overline{x}) \quad \text{and} \quad q = f_v(\overline{x}, \overline{x}).$$

The local stability character of $\overline{x}$ is now described by the linearized stability Theorem 1.1.1.

For the zero equilibrium of Eq(2.1) we have

$$p = \frac{\beta}{A} \quad \text{and} \quad q = \frac{\gamma}{A}$$

and so the linearized equation associated with Eq(2.1) about the zero equilibrium point is

$$z_{n+1} - \frac{\beta}{A} z_n - \frac{\gamma}{A} z_{n-1} = 0, \quad n = 0, 1, \ldots .$$

For the stability of the zero equilibrium of Eq(2.1), in addition to the local stability Theorem 1.1.1, we have the luxury of the global stability Theorem 1.3.1. As a consequence of these two theorems we obtain the following global result:

Theorem 2.2.1 *Assume that $A > 0$. Then the zero equilibrium point of the equation*

$$x_{n+1} = \frac{\beta x_n + \gamma x_{n-1}}{A + Bx_n + Cx_{n-1}}, \quad n = 0, 1, \ldots$$

is globally asymptotically stable when

$$\beta + \gamma \leq A$$

and is unstable when

$$\beta + \gamma > A.$$

Furthermore the zero equilibrium point is

a sink when	$A > \beta + \gamma$
a saddle point when	$A < \beta + \gamma < A + 2\beta$
a repeller when	$\beta + \gamma > A + 2\beta.$

2.3 Local Stability of the Positive Equilibrium

As we mentioned in Section 2.1, Eq(2.1) has a positive equilibrium when either

$$\alpha = 0 \quad \text{and} \quad \beta + \gamma > A$$

or

$$\alpha > 0.$$

In these cases the positive equilibrium $\overline{x}$ is unique, it satisfies Eqs(2.3) and (2.4), and it is given by (2.8). By using the identity

$$(B + C)\overline{x}^2 = (\beta + \gamma - A)\overline{x} + \alpha,$$

we see that

$$p = f_u(\overline{x}, \overline{x}) = \frac{(\beta A - \alpha B) + (\beta C - \gamma B)\overline{x}}{(A + (B + C)\overline{x})^2} = \frac{(\beta A - \alpha B) + (\beta C - \gamma B)\overline{x}}{A^2 + \alpha(B + C) + (B + C)(A + \beta + \gamma)\overline{x}}$$

and

$$q = f_v(\overline{x}, \overline{x}) = \frac{(\gamma A - \alpha C) - (\beta C - \gamma B)\overline{x}}{(A + (B + C)\overline{x})^2} = \frac{(\gamma A - \alpha C) - (\beta C - \gamma B)\overline{x}}{A^2 + \alpha(B + C) + (B + C)(A + \beta + \gamma)\overline{x}}.$$

Hence the conditions for the local asymptotic stability of $\overline{x}$ are (see Theorem 1.1.1) the following:

$$|p| < 1 - q \tag{2.11}$$

and

$$q > -1. \tag{2.12}$$

Inequalities (2.11) and (2.12) are equivalent to the following three inequalities:

$$\left[A + \beta + \gamma + 2\frac{\beta C - \gamma B}{B + C}\right]\overline{x} > -\alpha - \frac{A^2 + (\beta A - \alpha B) - (\gamma A - \alpha C)}{B + C} \tag{2.13}$$

$$(A + \beta + \gamma)\overline{x} > -\alpha - \frac{A^2 - (\beta A - \alpha B) - (\gamma A - \alpha C)}{B + C} \tag{2.14}$$

$$\left[A + \beta + \gamma - \frac{\beta C - \gamma B}{B + C}\right]\overline{x} > -\alpha - \frac{A^2 + (\gamma A - \alpha C)}{B + C}. \tag{2.15}$$

Out of these inequalities we will obtain explicit stability conditions by using the following procedure:

First we observe that Inequalities (2.13)-(2.15), are equivalent to some inequalities of the form

$$\overline{x} > \rho$$

and/or

$$\overline{x} < \sigma$$

for some $\rho, \sigma \in [0, \infty)$.

Now if we set

$$F(u) = (B + C)u^2 - (\beta + \gamma - A)u - \alpha,$$

it is clear that

$$F(\overline{x}) = 0$$

and that

$$\overline{x} > \rho \quad \text{if and only if} \quad F(\rho) < 0$$

while

$$\overline{x} < \sigma \quad \text{if and only if} \quad F(\sigma) > 0.$$

2.4 When is Every Solution Periodic with the Same Period?

The following four special examples of Eq(2.1)

$$x_{n+1} = \frac{1}{x_n}, \quad n = 0, 1, \ldots \tag{2.16}$$

$$x_{n+1} = \frac{1}{x_{n-1}}, \quad n = 0, 1, \ldots \tag{2.17}$$

$$x_{n+1} = \frac{1 + x_n}{x_{n-1}}, \quad n = 0, 1, \ldots \tag{2.18}$$

$$x_{n+1} = \frac{x_n}{x_{n-1}}, \quad n = 0, 1, \ldots \tag{2.19}$$

are remarkable in the sense that all positive solutions of each of these four nontrivial equations are periodic with periods $2, 4, 5$, and 6 respectively.

The following result characterizes all possible special cases of equations of the form of Eq(2.1) with the property that every solution of the equation is periodic with the same period. (See also [75], [73], and [74].)

Theorem 2.4.1 *Let $p \geq 2$ be a positive integer and assume that every positive solution of Eq(2.1) is periodic with period p. Then the following statements are true:*

(i) Assume $C > 0$. Then $A = B = \gamma = 0$.

(ii) Assume $C = 0$. Then $\gamma(\alpha + \beta) = 0$.

Proof. Consider the solution with

$$x_{-1} = 1 \quad \text{and} \quad x_0 \in (0, \infty).$$

Then clearly

$$x_0 = x_p \quad \text{and} \quad x_{p-1} = x_{-1} = 1$$

and so from Eq(2.1)

$$x_p = \frac{\alpha + \beta + \gamma x_{p-2}}{A + B + C x_{p-2}} = x_0.$$

Hence

$$(A + B)x_0 + (Cx_0 - \gamma)x_{p-2} = \alpha + \beta. \tag{2.20}$$

(i) Assume that $C > 0$. Then we claim that

$$A = B = 0. \tag{2.21}$$

Otherwise $A + B > 0$ and by choosing

$$x_0 > \max\left\{\frac{\alpha + \beta}{A + B}, \frac{\gamma}{C}\right\}$$

we see that (2.20) is impossible. Hence (2.21) holds. In addition to (2.21) we now claim that also

$$\gamma = 0. \tag{2.22}$$

Otherwise $\gamma > 0$ and by choosing

$$x_0 < \min\left\{\frac{\alpha + \beta}{A + B}, \frac{\gamma}{C}\right\}$$

we see that (2.20) is impossible. Thus (2.22) holds .

(ii) Assume $C = 0$ and, for the sake of contradiction, assume that

$$\gamma(\alpha + \beta) > 0.$$

Then by choosing x_0 sufficiently small, we see that (2.20) is impossible.

The proof is complete. □

Corollary 2.4.1 *Let $p \in \{2, 3, 4, 5, 6\}$. Assume that every positive solution of Eq(2.1) is periodic with period p. Then up to a change of variables of the form*

$$x_n = \lambda y_n$$

Eq(2.1) reduces to one of the equations (2.16)-(2.19) .

2.5 Existence of Prime Period-Two Solutions

In this section we give necessary and sufficient conditions for Eq(2.1) to have a prime period-two solution and we exhibit all prime period-two solutions of the equation. (See [50].)

Furthermore we are interested in conditions under which every solution of Eq(2.1) converges to a period-two solution in a nontrivial way according to the following convention.

Convention *Throughout this book when we say that "every solution of a certain difference equation converges to a periodic solution with period p" we mean that every solution converges to a, not necessarily prime, solution with period p and furthermore the set of solutions of the equation with prime period p is nonempty.*

Assume that

$$\dots, \phi, \psi, \phi, \psi, \dots$$

is a prime period-two solution of Eq(2.1). Then

$$\phi = \frac{\alpha + \beta\psi + \gamma\phi}{A + B\psi + C\phi}$$

and

$$\psi = \frac{\alpha + \beta\phi + \gamma\psi}{A + B\phi + C\psi}$$

from which we find that

$$C(\phi + \psi) = \gamma - \beta - A \tag{2.23}$$

and when

$$C > 0 \tag{2.24}$$

we also find that

$$(B - C)\phi\psi = \frac{\alpha C + \beta(\gamma - \beta - A)}{C}. \tag{2.25}$$

One can show that when

$$C = 0 \quad \text{and} \quad B > 0 \tag{2.26}$$

Eq(2.1) has a prime period-two solution if and only if

$$\gamma = \beta + A. \tag{2.27}$$

Furthermore when (2.26) and (2.27) hold

$$\dots, \phi, \psi, \phi, \psi, \dots$$

is a prime period-two solution of Eq(2.1) if and only if

$$\alpha + \beta(\phi + \psi) = B\phi\psi \quad \text{with} \quad \phi, \psi \in [0, \infty) \quad \text{and} \quad \phi \neq \psi$$

or equivalently

$$\phi > \frac{\beta}{B}, \quad \phi \neq \frac{\beta + \sqrt{\beta^2 + \alpha B}}{B}, \quad \text{and} \quad \psi = \frac{\alpha + \beta\phi}{-\beta + B\phi}. \tag{2.28}$$

Next assume that (2.24) holds and that Eq(2.1) possesses a prime period-two solution

$$\ldots, \phi, \psi, \phi, \psi, \ldots .$$

Then it follows from (2.23) and (2.25) that one of the following three conditions must be satisfied:

$$\gamma > \beta + A \quad \text{and} \quad \begin{array}{lll} B \neq C, \alpha = \beta = 0, & \text{and} & \phi\psi = 0; \quad (2.29) \\ B = C, \alpha = \beta = 0, & \text{and} & \phi\psi \geq 0; \quad (2.30) \\ B > C, \alpha + \beta > 0, & \text{and} & \phi\psi > 0. \quad (2.31) \end{array}$$

When (2.29) holds, the two cycle is

$$\ldots, 0, \frac{\gamma - A}{C}, 0, \frac{\gamma - A}{C}, \ldots .$$

When (2.30) holds, the two cycle is

$$\ldots, \phi, \frac{\gamma - A}{C} - \phi, \phi, \frac{\gamma - A}{C} - \phi, \ldots \quad \text{with} \quad 0 \leq \phi < \frac{\gamma - A}{C}.$$

Finally when (2.31) holds, the values ϕ and ψ of the two cycle are given by the two roots of the quadratic equation

$$t^2 - \frac{\gamma - \beta - A}{C} t + \frac{\alpha C + \beta(\gamma - \beta - A)}{C(B - C)} = 0$$

and we should also require that the discriminant of this quadratic equation be positive. That is,

$$\alpha < \frac{(\gamma - \beta - A)[B(\gamma - \beta - A) - C(\gamma + 3\beta - A)]}{4C^2}. \tag{2.32}$$

2.6 Local Asymptotic Stability of a Two Cycle

Let

$$\ldots, \phi, \psi, \phi, \psi, \ldots$$

be a two cycle of the difference equation (2.1).

Set

$$u_n = x_{n-1} \quad \text{and} \quad v_n = x_n \quad \text{for } n = 0, 1, \ldots$$

and write Eq(2.1) in the equivalent form:

$$\begin{aligned} u_{n+1} &= v_n \\ v_{n+1} &= \tfrac{\alpha+\beta v_n+\gamma u_n}{A+Bv_n+Cu_n}, \quad n = 0, 1, \ldots. \end{aligned}$$

Let T be the function on $[0,\infty) \times [0,\infty)$ defined by:

$$T\begin{pmatrix} u \\ v \end{pmatrix} = \begin{pmatrix} v \\ \frac{\beta v+\gamma u+\alpha}{Bv+Cu+A} \end{pmatrix}.$$

Then

$$\begin{pmatrix} \phi \\ \psi \end{pmatrix}$$

is a fixed point of T^2, the second iterate of T. Furthermore

$$T^2\begin{pmatrix} u \\ v \end{pmatrix} = \begin{pmatrix} g(u,v) \\ h(u,v) \end{pmatrix}$$

where

$$g(u,v) = \frac{\alpha+\beta v+\gamma u}{A+Bv+Cu} \quad \text{and} \quad h(u,v) = \frac{\alpha+\beta\frac{\alpha+\beta v+\gamma u}{A+Bv+Cu}+\gamma v}{A+B\frac{\alpha+\beta v+\gamma u}{A+Bv+Cu}+Cv}.$$

The two cycle is locally asymptotically stable if the eigenvalues of the Jacobian matrix J_{T^2}, evaluated at $\begin{pmatrix} \phi \\ \psi \end{pmatrix}$ lie inside the unit disk.

By definition

$$J_{T^2}\begin{pmatrix} \phi \\ \psi \end{pmatrix} = \begin{pmatrix} \frac{\partial g}{\partial u}(\phi,\psi) & \frac{\partial g}{\partial v}(\phi,\psi) \\ \frac{\partial h}{\partial u}(\phi,\psi) & \frac{\partial h}{\partial v}(\phi,\psi) \end{pmatrix}.$$

By computing the partial derivatives of g and h and by using the fact that

$$\phi = \frac{\alpha+\beta\psi+\gamma\phi}{A+B\psi+C\phi} \quad \text{and} \quad \psi = \frac{\alpha+\beta\phi+\gamma\psi}{A+B\phi+C\psi}$$

we find the following identities:

$$\begin{aligned} \frac{\partial g}{\partial u}(\phi,\psi) &= \frac{(\gamma A-\alpha C)+(\gamma B-\beta C)\psi}{(A+B\psi+C\phi)^2}, \\ \frac{\partial g}{\partial v}(\phi,\psi) &= \frac{(\beta A-\alpha B)-(\gamma B-\beta C)\phi}{(A+B\psi+C\phi)^2}, \\ \frac{\partial h}{\partial u}(\phi,\psi) &= \frac{[(\beta A-\alpha B)-(\gamma B-\beta C)\psi][(\gamma A-\alpha C)+(\gamma B-\beta C)\psi]}{(A+B\phi+C\psi)^2(A+B\psi+C\phi)^2}, \end{aligned}$$

$$\frac{\partial h}{\partial v}(\phi,\psi) = \frac{[(\beta A-\alpha B)-(\gamma B-\beta C)\psi][(\beta A-\alpha B)-(\gamma B-\beta C)\phi]}{(A+B\phi+C\psi)^2(A+B\psi+C\phi)^2} + \frac{(\gamma A-\alpha C)+(\gamma B-\beta C)\phi}{(A+B\phi+C\psi)^2}.$$

Set

$$\mathcal{S} = \frac{\partial g}{\partial u}(\phi,\psi) + \frac{\partial h}{\partial v}(\phi,\psi)$$

and

$$\mathcal{D} = \frac{\partial g}{\partial u}(\phi,\psi)\frac{\partial h}{\partial v}(\phi,\psi) - \frac{\partial g}{\partial v}(\phi,\psi)\frac{\partial h}{\partial u}(\phi,\psi).$$

Then it follows from Theorem 1.1.1 (c) that both eigenvalues of $J_{T^2}\begin{pmatrix}\phi\\ \psi\end{pmatrix}$ lie inside the unit disk $|\lambda| < 1$, if and only if

$$|\mathcal{S}| < 1 + \mathcal{D} \quad \text{and} \quad \mathcal{D} < 1.$$

See Sections 6.6, 6.9, and 7.4 for some applications.

Definition 2.6.1 *(Basin of Attraction of a Two Cycle)*
Let

$$\ldots, \phi, \psi, \phi, \psi, \ldots$$

be a two cycle of Eq(2.1). The **basin of attraction** *of this two cycle is the set B of all initial conditions (x_{-1}, x_0) through which the solution of Eq(2.1) converges to the two cycle.*

2.7 Convergence to Period-Two Solutions When $C = 0$

In this section we consider the equation

$$x_{n+1} = \frac{\alpha + \beta x_n + \gamma x_{n-1}}{A + Bx_n}, \quad n = 0, 1, \ldots \tag{2.33}$$

where

$$\alpha, \beta, \gamma, A, B \in [0, \infty) \quad \text{with} \quad \alpha + \beta + \gamma, A + B \in (0, \infty)$$

and nonnegative initial conditions and obtain a **signum invariant** for its solutions which will be used in the sequel (See Sections 4.7, 6.5, 6.7, and 10.2) to obtain conditions on α, β, γ, A and B so that every solution of Eq(2.33) converges to a period-two solution. That is, every solution converges to a (not necessarily prime) period-two solution and there exist prime period-two solutions. (See also Section 10.2 where we present the complete character of solutions of Eq(2.33).)

Now before we see that when $B > 0$ every solution of Eq(2.33) converges to a period-two solution if and only if (2.27) holds, we need the following result, the proof of which follows by straightforward computations.

Lemma 2.7.1 *Assume that (2.27) holds, and let* $\{x_n\}_{n=-1}^{\infty}$ *be a solution of Eq(2.33). Set*

$$J_n = \alpha + \beta(x_{n-1} + x_n) - Bx_{n-1}x_n \quad \text{for } n \geq 0. \tag{2.34}$$

Then the following statements are true:

(a)

$$J_{n+1} = \frac{\beta + A}{A + Bx_n} J_n \quad \text{for } n \geq 0.$$

In particular, the sign of the quantity J_n *is constant along each solution.*

(b)

$$x_{n+1} - x_{n-1} = \frac{J_n}{A + Bx_n} \quad \text{for } n \geq 0.$$

In particular, one and only one of the following three statements is true for each solution of Eq(2.33):

(i)

$$x_{n+1} - x_{n-1} = 0 \quad \text{for all} \quad n \geq 0;$$

(ii)

$$x_{n+1} - x_{n-1} < 0 \quad \text{for all} \quad n \geq 0;$$

(iii)

$$x_{n+1} - x_{n-1} > 0 \quad \text{for all} \quad n \geq 0.$$

Clearly *(i)* holds, if and only if, $(\beta + A)J_0 = 0$. In this case, the solution $\{x_n\}$ is periodic with period-two.

Statement *(ii)* holds if and only if $J_0 < 0$ and $\beta + A > 0$. In this case the subsequences $\{x_{2n}\}_{n=0}^{\infty}$ and $\{x_{2n-1}\}_{n=0}^{\infty}$ of the solution $\{x_n\}_{n=0}^{\infty}$ are both decreasing to finite limits and so in this case the solution converges to a period-two solution.

Finally *(iii)* holds, if and only if $J_0 > 0$ and $\beta + A > 0$. In this case the subsequences $\{x_{2n}\}_{n=0}^{\infty}$ and $\{x_{2n-1}\}_{n=0}^{\infty}$ of the solution $\{x_n\}_{n=0}^{\infty}$ are both increasing and so if we can show that they are bounded, the solution would converge to a period-two solution.

Now observe that

$$x_{n+1} - x_{n-1} = \frac{\beta + A}{A + Bx_n}(x_n - x_{n-2}) \quad \text{for} \quad n \geq 1$$

and so

$$x_{n+1} - x_{n-1} = (x_1 - x_{-1}) \prod_{k=1}^{n} \frac{\beta + A}{A + Bx_k}. \tag{2.35}$$

We will now employ (2.35) to show that the subsequences of even and odd terms of the solution are bounded. We will give the details for the subsequence of even terms. The proof for the subsequence of odd terms is similar and will be omitted. To this end assume, for the sake of contradiction, that the subsequence of even terms increases to ∞. Then there exists $N \geq 0$ such that

$$\rho = \frac{A+\beta}{A+Bx_{2N+1}} \frac{A+\beta}{A+Bx_{2N}} < 1.$$

Clearly for all $n > N$,

$$\frac{A+\beta}{A+Bx_{2n+1}} \frac{A+\beta}{A+Bx_{2n}} < \frac{A+\beta}{A+Bx_{2N+1}} \frac{A+\beta}{A+Bx_{2N}} = \rho.$$

Set

$$K = |x_{2N} - x_{2N-2}|.$$

Then

$$|x_{2N+2} - x_{2N}| = \frac{A+\beta}{A+Bx_{2N+1}} \frac{A+\beta}{A+Bx_{2N}} |x_{2N} - x_{2N-2}| < K\rho$$

$$|x_{2N+4} - x_{2N+2}| = \frac{A+\beta}{A+Bx_{2N+3}} \frac{A+\beta}{A+Bx_{2N+2}} |x_{2N+2} - x_{2N}| < K\rho^2$$

and by induction

$$|x_{2N+2m} - x_{2N+2(m-1)}| < K\rho^m \quad \text{for} \quad m = 1, 2, \ldots .$$

But then

$$\begin{aligned} x_{2N+2m} &\leq |x_{2N+2m} - x_{2N+2m-2}| + \ldots + |x_{2N+2} - x_{2N}| + x_{2N} \\ &\leq K \sum_{i=1}^{m} \rho^i + x_{2N} < \frac{K\rho}{1-\rho} + x_{2N} \end{aligned}$$

which contradicts the hypothesis that the subsequence of even terms is unbounded.

In summary we have established the following result.

Theorem 2.7.1 *Assume $B > 0$. Then the following statements hold:*

(a) Eq(2.33) has a prime period-two solution if and only if (2.27) holds.

(b) When (2.27) holds all prime period-two solutions

$$\ldots, \phi, \psi, \phi, \psi, \ldots$$

of Eq(2.33) are given by (2.28).

(c) When (2.27) holds every solution of Eq(2.33) converges to a period-two solution.

2.8 Invariant Intervals

An **invariant interval** for Eq(2.1) is an interval I with the property that if two consecutive terms of the solution fall in I then all the subsequent terms of the solution also belong to I. In other words I is an invariant interval for the difference equation

$$x_{n+1} = f(x_n, x_{n-1}), \quad n = 0, 1, \ldots \tag{2.36}$$

if $x_{N-1}, x_N \in I$ for some $N \geq 0$, then $x_n \in I$ for every $n > N$.

Invariant intervals for Eq(2.36) are determined from the intervals where the function $f(x, y)$ is monotonic in its arguments.

For Eq(2.1) the partial derivatives are given by (2.9) and (2.10) and are

$$f_x(x,y) = \frac{L + My}{(A + Bx + Cy)^2} \quad \text{and} \quad f_y(x,y) = \frac{N - Mx}{(A + Bx + Cy)^2},$$

where

$$L = \beta A - \alpha B, \quad M = \beta C - \gamma B, \quad \text{and} \quad N = \gamma A - \alpha C.$$

The following table gives the signs of f_x and f_y in all possible nondegenerate cases.

Case	L	M	N	Signs of derivatives f_x and f_y
1	+	+	+	$f_x > 0$ $f_y > 0$ if $x < \frac{N}{M}$; $f_y < 0$ if $x > \frac{N}{M}$
2	+	+	0	$f_x > 0$ and $f_y < 0$
3	+	+	−	$f_x > 0$ and $f_y < 0$
4	+	0	+	$f_x > 0$ and $f_y > 0$
5	+	0	0	$f_x > 0$ and $f_y = 0$
6	+	−	+	$f_x > 0$ if $y < -\frac{L}{M}$; $f_x < 0$ if $y > -\frac{L}{M}$ $f_y > 0$
7	0	+	0	$f_x > 0$ and $f_y < 0$
8	0	+	−	$f_x > 0$ and $f_y < 0$
9	0	−	+	$f_x < 0$ and $f_y > 0$
10	0	−	0	$f_x < 0$ and $f_y > 0$
11	−	+	−	$f_x > 0$ if $y > -\frac{L}{M}$; $f_x < 0$ if $y < -\frac{L}{M}$ $f_y < 0$
12	−	0	0	$f_x < 0$ and $f_y = 0$
13	−	0	−	$f_x < 0$ and $f_y < 0$
14	−	−	+	$f_x < 0$ and $f_y > 0$
15	−	−	0	$f_x < 0$ and $f_y > 0$
16	−	−	−	$f_x < 0$ $f_y > 0$ if $x > \frac{N}{M}$; $f_y < 0$ if $x < \frac{N}{M}$

Using this table, we obtain the following table of invariant intervals for Eq(2.1).

Case	Invariant intervals
1	$[0, \frac{N}{M}]$ if $\frac{\alpha M+(\beta+\gamma)N}{AM+(B+C)N} < \frac{N}{M}$ $[\frac{N}{M}, \frac{(\beta-A)M-CN+\sqrt{((\beta-A)M-NC)^2+4BM(\alpha M+\gamma N)}}{2BM}]$ if $\frac{\alpha M+(\beta+\gamma)N}{AM+(B+C)N} > \frac{N}{M}$ and $B>0$ $[\frac{N}{M}, \frac{\alpha M+\gamma N}{(A-\beta)M+CN}]$ if $B=0$ and $(A-\beta)M+CN>0$
2	$[0, \frac{\beta-A+\sqrt{(\beta-A)^2+4B\alpha}}{2B}]$ if $B>0$ $[0, \frac{\alpha}{A-\beta}]$ if $B=0, \alpha>0$ and $A>\beta$
3	$[0, \frac{\beta-A+\sqrt{(\beta-A)^2+4B\alpha}}{2B}]$ if $B>0$ $[0, \frac{\alpha}{A-\beta}]$ if $B=0, \alpha>0$ and $A>\beta$
4	$[0, \frac{\beta+\gamma-A+\sqrt{(\beta+\gamma-A)^2+4\alpha(B+C)}}{2(B+C)}]$ if $B+C>0$
5	$[\frac{\alpha}{A}, \frac{\beta-A+\sqrt{(\beta-A)^2+4\alpha B}}{2B}]$ if $B>0$ $[\frac{\alpha}{A}, \frac{\alpha}{A-\beta}]$ if $B=0$ and $A>\beta$
6	$[0, -\frac{L}{M}]$ if $\frac{\alpha M-(\beta+\gamma)L}{AM-(B+C)L} < -\frac{L}{M}$ $[-\frac{L}{M}, \frac{(A-\gamma)M-BL+\sqrt{((A-\gamma)M-BL)^2+4CM(\alpha M-\beta L)}}{-2CM}]$ if $C\neq 0$ and $\frac{\beta MK+\alpha M-\gamma L}{BMK+AM-CL} > -\frac{L}{M}$ $[-\frac{L}{M}, \frac{\beta L-\alpha M}{(\gamma-A)M+BL}]$ if $C=0, (\gamma-A)M+BL>0$ and $-(\beta M+BL)K \geq \alpha M+AL$
7	$[\frac{\gamma}{C}, \frac{\beta}{B}]$ if $B>0$ $[\frac{\gamma}{C}, \frac{\gamma^2}{C(\gamma-\beta)}]$ if $B=0$ and $\beta<\gamma$
8	$[0, \frac{\beta}{B}]$ if $B\neq 0$ $[\frac{\gamma}{C}, \frac{\alpha C+\gamma^2}{\gamma-\beta C}]$ if $B=0$ and $\gamma>\beta C$
9	$[0, \frac{\gamma}{C}]$ if $C>0$ $[k, K]$ if $C=0$ where k and K satisfy $\alpha+\beta k+(\gamma-A)K \leq BkK \leq \alpha+\beta K-Ak$
10	$[\frac{\beta}{B}, \frac{\gamma}{C}]$ if $C>0$ $[\frac{\beta}{B}, \frac{\beta^2}{B(\beta-\gamma)}]$ if $C=0$ and $\beta>\gamma$
11	$[\frac{\alpha M-L(\beta+\gamma)}{AM-L(B+C)}, -\frac{L}{M}]$ if $\frac{\alpha M-L(\beta+\gamma)}{AM-L(B+C)} < -\frac{L}{M}$ $[-\frac{L}{M}, \frac{(\beta-A)M+LC+\sqrt{((\beta-A)M+LC)^2+4BM(\alpha M-\gamma L)}}{2BM}]$ if $\frac{\alpha M-\beta L+\gamma MK}{AM-BL+CMK} > -\frac{L}{M}$
12	$[\frac{\alpha(A+\beta)}{A^2+B\alpha}, \frac{\alpha}{A}]$ if $A>0$ $[\frac{\beta}{B}, \frac{\beta}{B}+\frac{\alpha}{\beta}]$ if $A=0$ and $\beta>0$
13	$[0, \frac{\alpha}{A}]$ if $A>0$ $[k, K]$ if $A=0$ where k and K satisfy $\frac{\alpha+(\beta+\gamma)k}{B+C} \leq kK \leq \frac{\alpha+(\beta+\gamma)K}{B+C}$
14	$[0, \frac{\gamma}{C}]$ if $C>0$ $[k, K]$ if $C=0$ where k and K satisfy $\alpha+\beta k+(\gamma-A)K \leq BkK \leq \alpha+\beta K-Ak$
15	$[0, \frac{\gamma}{C}]$ if $C>0$ $[\frac{\beta}{B}, \frac{\alpha B+\beta^2}{B(\beta-\gamma)}]$ if $A=0$ and $\beta>\gamma$
16	$[\frac{\alpha M+(\beta+\gamma)N}{AM+(B+C)N}, \frac{N}{M}]$ if $\frac{\alpha M+(\beta+\gamma)N}{AM+(B+C)N} < \frac{N}{M}$ $[\frac{N}{M}, \frac{(\gamma-A)M-BN+\sqrt{((\gamma-A)M-BN)^2+4BM(\alpha M+\beta N)}}{2CM}]$ if $\frac{\alpha M+\beta K+\gamma NM}{AM+BK+CNM} > \frac{N}{M}$

2.9 Open Problems and Conjectures

What is it that makes all solutions of a nonlinear difference equation periodic with the same period?

For a linear equation, every solution is periodic with period $p \geq 2$, if and only if every root of the characteristic equation is a pth root of unity.

Open Problem 2.9.1 *Assume that $f \in C^1[(0,\infty) \times (0,\infty), (0,\infty)]$ is such that every positive solution of the equation*

$$x_{n+1} = f(x_n, x_{n-1}), \quad n = 0, 1, \ldots \tag{2.37}$$

is periodic with period $p \geq 2$.

Is it true that the linearized equation about a positive equilibrium point of Eq(2.37) has the property that every one of its solutions is also periodic with the same period p?

What is it that makes every positive solution of a difference equation converge to a periodic solution with period $p \geq 2$?

Open Problem 2.9.2 *Let $f \in C^1[[0,\infty) \times [0,\infty), [0,\infty)]$ and let $p \geq 2$ be an integer. Find necessary and sufficient conditions so that every nonnegative solution of the difference equation (2.37) converges to a periodic solution with period p. In particular address the case $p = 2$.*

Open Problem 2.9.3 *Let $p \geq 2$ be an integer and assume that every positive solution of Eq(2.1) with*

$$\alpha, \beta, \gamma, A, B, C \in [0,\infty)$$

converges to a solution with period p . Is it true that

$$p \in \{2, 4, 5, 6\}?$$

Open Problem 2.9.4 *It is known (see Sections 2.7, 4.7, 6.5, and 6.7) that every positive solution of each of the following three equations*

$$x_{n+1} = 1 + \frac{x_{n-1}}{x_n}, \quad n = 0, 1, \ldots \tag{2.38}$$

$$x_{n+1} = \frac{1 + x_{n-1}}{1 + x_n}, \quad n = 0, 1, \ldots \tag{2.39}$$

$$x_{n+1} = \frac{x_n + 2x_{n-1}}{1 + x_n}, \quad n = 0, 1, \ldots \tag{2.40}$$

converges to a solution with (not necessarily prime) period-two:

$$\ldots, \phi, \psi, \phi, \psi, \ldots \quad . \tag{2.41}$$

In each case, determine ϕ and ψ in terms of the initial conditions x_{-1} and x_0.

Conversely, if (2.41) is a period-two solution of Eq(2.38) or Eq(2.39) or Eq(2.40), determine all initial conditions $(x_{-1}, x_0) \in (0, \infty) \times (0, \infty)$ for which the solution $\{x_n\}_{n=-1}^{\infty}$ converges to the period-two solution (2.41).

Conjecture 2.9.1 *Assume*

$$\alpha, \beta, \gamma, A, B, C \in (0, \infty).$$

Show that every positive solution of Eq(2.1) is bounded.

Open Problem 2.9.5 *Assume*

$$\alpha, \beta, \gamma, A, B \in (0, \infty) \quad \text{and} \quad \gamma > \beta + A.$$

Determine the set of initial conditions x_{-1} and x_0 for which the solution $\{x_n\}_{n=-1}^{\infty}$ of Eq(2.33) is bounded.

(See Sections 2.7, 4.7, 6.5, 6.7, and 10.2.)

Conjecture 2.9.2 *Assume*

$$\alpha, \beta, \gamma, A, B, C \in (0, \infty)$$

and that Eq(2.1) has no period-two solution. Show that the positive equilibrium is globally asymptotically stable.

Conjecture 2.9.3 *Assume*

$$\alpha, \beta, \gamma, A, B, C \in (0, \infty)$$

and that Eq(2.1) has a positive prime period-two solution. Show that the positive equilibrium of Eq(2.1) is a saddle point.

Open Problem 2.9.6 *Assume*

$$\alpha, \beta, \gamma, A, B, C \in [0, \infty).$$

Obtain necessary and sufficient conditions in terms of the coefficients so that every positive solution of Eq(2.1) is bounded.

Conjecture 2.9.4 *Assume that*

$$\alpha, \beta, \gamma, A, B, C \in [0, \infty).$$

Show that Eq(2.1) cannot have a positive prime two cycle which attracts all positive non-equilibrium solutions.

Open Problem 2.9.7 *Assume* $f \in C[(0, \infty), (0, \infty)]$. *Obtain necessary and sufficient conditions on* f *for every solution of the difference equation*

$$x_{n+1} = \frac{f(x_n)}{x_{n-1}}, \quad n = 0, 1, \ldots$$

to be bounded.

(See Section 5.2).

Open Problem 2.9.8 *Assume* $f \in C[(0, \infty), (0, \infty)]$. *Obtain necessary and sufficient conditions on* f *for every positive solution of the equation*

$$x_{n+1} = 1 + \frac{f(x_{n-1})}{x_n}, \quad n = 0, 1, \ldots$$

to converge to a period-two solution.

(See Section 4.7).

Open Problem 2.9.9 *Extend Theorem 2.7.1 to third order difference equations*

$$x_{n+1} = \frac{\alpha + \beta x_n + \gamma x_{n-1} + \delta x_{n-2}}{A + Bx_{n-1} + Dx_{n-2}}, \quad n = 0, 1, \ldots$$

with nonnegative parameters and nonnegative initial conditions such that

$$\gamma = \beta + \delta + A.$$

What additional conditions should the parameters satisfy?

Open Problem 2.9.10 *Extend the linearized stability Theorem 1.1.1 to third order difference equations*

$$x_{n+1} = f(x_n, x_{n-1}, x_{n-2}), \quad n = 0, 1, \dots . \tag{2.42}$$

For the equation

$$\lambda^3 + a\lambda^2 + b\lambda + c = 0$$

one can see that all roots lie inside the unit disk $|\lambda| < 1$ *if and only if*

$$\left.\begin{array}{l} |a + c| < 1 + b \\ |a - 3c| < 3 - b \\ b + c^2 < 1 + ac \end{array}\right\}.$$

But how about necessary and sufficient conditions for the equilibrium of Eq(2.42) to be a repeller, or a saddle point, or nonhyperbolic?

Extend these results to fourth order difference equation

$$x_{n+1} = f(x_n, x_{n-1}, x_{n-2}, x_{n-3}), \quad n = 0, 1, \dots .$$

Chapter 3

$(1,1)$-Type Equations

3.1 Introduction

Eq(1) contains the following nine equations of the $(1,1)$-type:

$$x_{n+1} = \frac{\alpha}{A}, \quad n = 0, 1, \ldots \tag{3.1}$$

$$x_{n+1} = \frac{\alpha}{Bx_n}, \quad n = 0, 1, \ldots \tag{3.2}$$

$$x_{n+1} = \frac{\alpha}{Cx_{n-1}}, \quad n = 0, 1, \ldots \tag{3.3}$$

$$x_{n+1} = \frac{\beta x_n}{A}, \quad n = 0, 1, \ldots \tag{3.4}$$

$$x_{n+1} = \frac{\beta}{B}, \quad n = 0, 1, \ldots \tag{3.5}$$

$$x_{n+1} = \frac{\beta x_n}{Cx_{n-1}}, \quad n = 0, 1, \ldots \tag{3.6}$$

$$x_{n+1} = \frac{\gamma x_{n-1}}{A}, \quad n = 0, 1, \ldots \tag{3.7}$$

$$x_{n+1} = \frac{\gamma x_{n-1}}{Bx_n}, \quad n = 0, 1, \ldots \tag{3.8}$$

and

$$x_{n+1} = \frac{\gamma}{C}, \quad n = 0, 1, \ldots . \tag{3.9}$$

Please recall our classification convention in which all parameters that appear in these equations are positive, the initial conditions are nonnegative, and the denominators are always positive.

Of these nine equations, Eqs(3.1), (3.5), and (3.9) are trivial. Eqs (3.4) and (3.7) are linear difference equation. Every solution of Eq(3.2) is periodic with period two and every solution of Eq(3.4) is periodic with period four. The remaining two equations, namely, Eqs(3.6) and (3.8), are treated in the next two sections.

3.2 The Case $\alpha = \gamma = A = B = 0: \ x_{n+1} = \frac{\beta x_n}{Cx_{n-1}}$

This is the $(1,1)$-type Eq(3.6) which by the change of variables

$$x_n = \frac{\beta}{C} y_n$$

reduces to the equation

$$y_{n+1} = \frac{y_n}{y_{n-1}}, \quad n = 0, 1, \ldots . \tag{3.10}$$

It is easy to see that every positive solution of Eq(3.10) is periodic with period six. Indeed the solution with initial conditions y_{-1} and y_0 is the six-cycle:

$$y_{-1}, y_0, \frac{y_0}{y_{-1}}, \frac{1}{y_{-1}}, \frac{1}{y_0}, \frac{y_{-1}}{y_0}, \ldots .$$

It is interesting to note that except for the equilibrium solution $\bar{y} = 1$, every other solution of Eq(3.10) has prime period six.

3.3 The Case $\alpha = \beta = A = C = 0:\ x_{n+1} = \frac{\gamma x_{n-1}}{Bx_n}$

This is the $(1,1)$-type Eq(3.8) which by the change of variables

$$x_n = \frac{\gamma}{B} e^{y_n}$$

reduces to the linear equation

$$y_{n+1} + y_n - y_{n-1} = 0, \quad n = 0, 1, \ldots . \tag{3.11}$$

The general solution of Eq(3.11) is

$$y_n = C_1 \left(\frac{-1+\sqrt{5}}{2} \right)^n + C_2 \left(\frac{-1-\sqrt{5}}{2} \right)^n, \quad n = 0, 1, \ldots$$

from which the behavior of solutions is easily derived.

3.4 Open Problems and Conjectures

Of the nine equations in this chapter we would like to single out the two nonlinear equations, (3.2) and (3.6), which possess the property that every positive nontrivial solution of each of these two equations is periodic with the same prime period, namely two and six, respectively. Eq(3.3) also has the property that every nontrivial positive solution is periodic with prime period four but this equation is really a variant of Eq(3.2). Also every solution of Eq(3.7) is periodic with period two but this is a linear equation.

Eq(3.2) is of the form

$$x_{n+1} = f(x_n), \quad n = 0, 1, \ldots \tag{3.12}$$

where

$$f \in C[(0, \infty), (0, \infty)], \tag{3.13}$$

while Eq(3.6) is of the form

$$x_{n+1} = f(x_n, x_{n-1}), \quad n = 0, 1, \ldots \tag{3.14}$$

where

$$f \in C[(0, \infty) \times (0, \infty), (0, \infty)]. \tag{3.15}$$

What is it that makes every solution of a nonlinear difference equation periodic with the same period? We believe this is a question of paramount importance and so we pose the following open problems.

Open Problem 3.4.1 *Let $k \geq 2$ be a positive integer and assume that (3.13) holds. Obtain necessary and sufficient conditions on f so that every positive solution of Eq(3.12) is periodic with period k.*

Open Problem 3.4.2 *Let $k \geq 2$ be a positive integer and assume that (3.15) holds. Obtain necessary and sufficient conditions on f so that every positive solution of Eq(3.14) is periodic with period k. In particular address the cases*

$$k = 4, 5, 6.$$

Open Problem 3.4.3 *Let k be a positive integer and assume that*

$$\alpha, \beta, \gamma \in (-\infty, \infty).$$

Obtain necessary and sufficient conditions on α, β, γ so that every solution of the equation

$$x_{n+1} = \alpha\,|x_n| + \beta x_{n-1} + \gamma, \quad n = 0, 1, \ldots$$

with real initial conditions is periodic with period k. In particular address the cases:

$$k = 5, 6, 7, 8, 9.$$

(See [10], [17], and [59].)

Conjecture 3.4.1 *Let $f \in C^1[[0, \infty), [0, \infty)]$ be such that every positive nontrivial solution of the difference equation*

$$x_{n+1} = \frac{f(x_n)}{x_{n-1}}, \quad n = 0, 1, \ldots \tag{3.16}$$

is periodic with prime period six. Show that

$$f(x) = x.$$

Conjecture 3.4.2 *Let $f \in C^1[[0, \infty), [0, \infty)]$ be such that every positive nontrivial solution of the difference equation (3.16) is periodic with prime period five. Show that*

$$f(x) = 1 + x.$$

Open Problem 3.4.4 *Obtain necessary and sufficient conditions on $f \in C[[0,\infty),[0,\infty)]$ such that every positive nontrivial solution of the equation*

$$x_{n+1} = \frac{f(x_n)}{x_{n-1}}, \quad n = 0, 1, \dots \tag{3.17}$$

is periodic with prime period three.

Open Problem 3.4.5 *Let $\{p_n\}_{n=0}^{\infty}$ be a sequence of nonzero real numbers. In each of the following cases investigate the asymptotic behavior of all positive solutions of the difference equation*

$$x_{n+1} = \frac{p_n x_n}{x_{n-1}}, \quad n = 0, 1, \dots . \tag{3.18}$$

(i) $\{p_n\}_{n=0}^{\infty}$ *converges to a finite limit;*

(ii) $\{p_n\}_{n=0}^{\infty}$ *converges to* $\pm\infty$*;*

(iii) $\{p_n\}_{n=0}^{\infty}$ *is periodic with prime period* $k \geq 2$.

Chapter 4

$(1, 2)$-Type Equations

4.1 Introduction

Eq(1) contains the following nine equations of the $(1, 2)$-type:

$$x_{n+1} = \frac{\alpha}{A + Bx_n}, \quad n = 0, 1, \ldots \tag{4.1}$$

$$x_{n+1} = \frac{\alpha}{A + Cx_{n-1}}, \quad n = 0, 1, \ldots \tag{4.2}$$

$$x_{n+1} = \frac{\alpha}{Bx_n + Cx_{n-1}}, \quad n = 0, 1, \ldots \tag{4.3}$$

$$x_{n+1} = \frac{\beta x_n}{A + Bx_n}, \quad n = 0, 1, \ldots \tag{4.4}$$

$$x_{n+1} = \frac{\beta x_n}{A + Cx_{n-1}}, \quad n = 0, 1, \ldots \tag{4.5}$$

$$x_{n+1} = \frac{\beta x_n}{Bx_n + Cx_{n-1}}, \quad n = 0, 1, \ldots \tag{4.6}$$

$$x_{n+1} = \frac{\gamma x_{n-1}}{A + Bx_n}, \quad n = 0, 1, \ldots \tag{4.7}$$

$$x_{n+1} = \frac{\gamma x_{n-1}}{A + Cx_{n-1}}, \quad n = 0, 1, \ldots \tag{4.8}$$

and

$$x_{n+1} = \frac{\gamma x_{n-1}}{Bx_n + Cx_{n-1}}, \quad n = 0, 1, \ldots . \tag{4.9}$$

Please recall our classification convention in which all parameters that appear in these equations are positive, the initial conditions are nonnegative, and the denominators are always positive.

Two of these equations, namely Eqs(4.1) and (4.4) are Riccati-type difference equations. (See Section 1.6 in the Preliminary Results). Also Eqs(4.2) and (4.8) are essentially Riccati equations. Indeed if $\{x_n\}$ is a solution of Eq(4.2), then the subsequences $\{x_{2n-1}\}$ and $\{x_{2n}\}$ satisfy the Riccati equation of the form of Eq(4.1). Similarly, if $\{x_n\}$ is a solution of Eq(4.8), then the subsequences $\{x_{2n-1}\}$ and $\{x_{2n}\}$ satisfy the Riccati equation

$$y_{n+1} = \frac{\gamma y_n}{A + Cy_n} \quad n = 0, 1, \ldots .$$

It is also interesting to note that the change of variables

$$x_n = \frac{1}{y_n}$$

reduces the Riccati equation (4.4) to the linear equation

$$y_{n+1} = \frac{A}{\beta} y_n + \frac{B}{\beta}, \quad n = 0, 1, \ldots$$

from which the global behavior of solutions is easily derived.

In view of the above, Eqs(4.2), (4.4), and (4.8) will not be discussed any further in this chapter.

4.2 The Case $\beta = \gamma = C = 0$: $x_{n+1} = \frac{\alpha}{A+Bx_n}$

This is the $(1, 2)$-type Eq(4.1) which by the change of variables

$$x_n = \frac{A}{B} y_n$$

reduces to the Riccati equation

$$y_{n+1} = \frac{p}{1 + y_n}, \qquad n = 0, 1, \ldots \tag{4.10}$$

where

$$p = \frac{\alpha B}{A^2}.$$

The Riccati number associated with this equation (see Section 1.6) is

$$R = -p < 0$$

and so the following result is a consequence of Theorem 1.6.2.

Theorem 4.2.1 *The positive equilibrium*

$$\bar{y} = \frac{-1 + \sqrt{1 + 4p}}{2}$$

of Eq(4.10) is globally asymptotically stable.

4.3 The Case $\beta = \gamma = A = 0: \; x_{n+1} = \frac{\alpha}{Bx_n + Cx_{n-1}}$

This is the $(1, 2)$-type Eq(4.3) which by the change of variables

$$x_n = \frac{\sqrt{\alpha}}{y_n} \tag{4.11}$$

is transformed to the difference equation

$$y_{n+1} = \frac{B}{y_n} + \frac{C}{y_{n-1}}, \qquad n = 0, 1, \ldots . \tag{4.12}$$

Eq(4.12) was investigated in [65]. (See also [12].)

We will first show that every solution of Eq(4.12) is bounded and then we will use the method of "limiting solutions" to show that every solution of Eq(4.12) converges to its equilibrium $\sqrt{B + C}$.

Theorem 4.3.1 *Every solution of Eq(4.12) is bounded and persists.*

Proof. It is easy to see that if a solution of Eq(4.12) is bounded from above then it is also bounded from below and vice-versa. So if the theorem is false there should exist a solution $\{y_n\}_{n=-1}^{\infty}$ which is neither bounded from above nor from below. That is,

$$\limsup_{n \to \infty} y_n = \infty \quad \text{and} \quad \liminf_{n \to \infty} y_n = 0.$$

Then clearly we can find indices i and j with $1 \leq i < j$ such that

$$y_i > y_n > y_j \quad \text{for all } n \in \{-1, \ldots, j-1\}.$$

Hence

$$y_j = \frac{B}{y_{j-1}} + \frac{C}{y_{j-2}} > \frac{B+C}{y_i}$$

and

$$y_i = \frac{B}{y_{i-1}} + \frac{C}{y_{i-2}} \leq \frac{B+C}{y_j}.$$

That is ,

$$B + C < y_i y_j \leq B + C$$

which is impossible. □

Theorem 4.3.2 *The equilibrium* $\overline{y} = \sqrt{B+C}$ *of Eq(4.12) is globally asymptotically stable.*

Proof. The linearized equation of Eq(4.12) about the equilibrium $\overline{y} = \sqrt{B+C}$ is

$$z_{n+1} = -\frac{B}{B+C} z_n - \frac{C}{B+C} z_{n-1}, \quad n = 0, 1, \ldots . \tag{4.13}$$

and by Theorem 1.1.1, $\overline{y}$ is locally asymptotically stable for all positive values of B and C.

It remains to be shown that $\overline{y}$ is a global attractor of all solutions of Eq(4.12). To this end, let $\{y_n\}_{n=-1}^{\infty}$ be a solution of Eq(4.12). Then by Theorem 4.3.1 , $\{y_n\}_{n=-1}^{\infty}$ is bounded and persists. Therefore there exist positive numbers m and M such that

$$m \leq y_n \leq M \quad \text{for } n \geq -1.$$

Let

$$I = \liminf_{n\to\infty} y_n \quad \text{and} \quad S = \limsup_{n\to\infty} y_n.$$

We will show that

$$I = S.$$

By the theory of limiting solutions (see Section 1.5) there exist two solutions $\{I_n\}_{n=-3}^{\infty}$ and $\{S_n\}_{n=-3}^{\infty}$ of the difference equation

$$y_{n+1} = \frac{B}{y_n} + \frac{C}{y_{n-1}}, \quad n = -3, -2, \ldots$$

with the following properties:

$$I_0 = I, \quad S_0 = S$$

and

$$I_n, S_n \in [I, S] \quad \text{for} \;\; n \geq -3.$$

Then

$$S = \frac{B}{S_{-1}} + \frac{C}{S_{-2}} \leq \frac{B+C}{I}$$

and

$$I = \frac{B}{I_{-1}} + \frac{C}{I_{-2}} \geq \frac{B+C}{S}.$$

Hence

$$B + C = SI.$$

Thus

$$\frac{B}{S_{-1}} + \frac{C}{S_{-2}} = S = \frac{B+C}{I} = \frac{B}{I} + \frac{C}{I}$$

from which it follows that

$$S_{-1} = S_{-2} = I. \tag{4.14}$$

Therefore

$$\frac{B}{S} + \frac{C}{S} = I = S_{-1} = \frac{B}{S_{-2}} + \frac{C}{S_{-3}}$$

from which it follows that

$$S = S_{-2}. \tag{4.15}$$

From (4.14) and (4.15) we conclude that

$$S = I$$

and the proof is complete. □

4.4 The Case $\alpha = \gamma = B = 0: \; x_{n+1} = \frac{\beta x_n}{A+Cx_{n-1}}$ - Pielou's Equation

This is the $(1, 2)$-type Eq(4.5) which by the change of variables

$$x_n = \frac{A}{C} y_n,$$

reduces to **Pielou's difference equation**

$$y_{n+1} = \frac{p y_n}{1 + y_{n-1}}, \quad n = 0, 1, \ldots \tag{4.16}$$

where

$$p = \frac{\beta}{A}.$$

This equation was proposed by Pielou in her books ([66], p.22) and ([67], p.79) as a discrete analogue of the **delay logistic equation**

$$\frac{dN}{dt} = rN(t)\left(1 - \frac{N(t-\tau)}{P}\right), \; t \geq 0$$

which is a prototype of modelling the dynamics of single-species.

Eq(4.16) was investigated in [55]. (See also ([42], p.75).)

When

$$p \leq 1,$$

it follows from Eq(4.16) that every positive solution converges to 0.

Furthermore, 0 is locally asymptotically stable when $p \leq 1$ and unstable when $p > 1$.

So the interesting case for Eq(4.16) is when

$$p > 1. \tag{4.17}$$

In this case the zero equilibrium of Eq(4.16) is unstable and Eq(4.16) possesses the unique positive equilibrium

$$\overline{y} = p - 1,$$

which is locally asymptotically stable.

In fact by employing Theorem 1.4.2, it follows that when (4.17) holds, $\overline{x}$ is globally asymptotically stable.

We summarize the above discussion in the following theorem.

Theorem 4.4.1 *(a) Assume*

$$p \leq 1.$$

Then the zero equilibrium of Eq(4.16) is globally asymptotically stable.

(b) Assume $y_0 \in (0, \infty)$ and

$$p > 1.$$

Then the positive equilibrium $\overline{y} = p - 1$ of Eq(4.16) is globally asymptotically stable.

4.5 The Case $\alpha = \gamma = A = 0: \; x_{n+1} = \frac{\beta x_n}{Bx_n + Cx_{n-1}}$

This is the (1,2)-type Eq(4.6) which by the change of variables

$$x_n = \frac{\beta}{B + Cy_n},$$

reduces to the difference equation

$$y_{n+1} = \frac{p + y_n}{p + y_{n-1}}, \quad n = 0, 1, \ldots \tag{4.18}$$

where

$$p = \frac{B}{C}.$$

Eq(4.18) was investigated in ([42] Corollary 3.4.1 (e), p. 73) where it was shown that its equilibrium one is globally asymptotically stable .

4.6 The Case $\alpha = \beta = C = 0: \; x_{n+1} = \frac{\gamma x_{n-1}}{A+Bx_n}$

This is the $(1,2)$-type Eq(4.7) which by the change of variables

$$x_n = \frac{\gamma}{B} y_n,$$

reduces to the difference equation

$$y_{n+1} = \frac{y_{n-1}}{p + y_n}, \qquad n = 0, 1, \ldots \tag{4.19}$$

where

$$p = \frac{A}{\gamma} \in (0, \infty).$$

Eq(4.19) was investigated in [28].

The following local result is a straightforward application of the linearized stability Theorem 1.1.1.

Lemma 4.6.1 *(a) Assume*

$$p > 1.$$

Then 0 is the only equilibrium point of Eq(4.19) and it is locally asymptotically stable.

(b) Assume

$$p < 1.$$

Then 0 and $\overline{y} = 1 - p$ are the only equilibrium points of Eq(4.19) and they are both unstable. In fact, 0 is a repeller and $\overline{y} = 1 - p$ is a saddle point equilibrium.

The oscillatory character of the solutions of Eq(4.19) is a consequence of Theorem 1.7.1. That is, except possibly for the first semicycle, every solution of Eq(4.19) has semicycles of length one.

It follows directly from Eq(4.19) (see also Section 2.5) that

$$y_{n+1} < \frac{1}{p} y_{n-1} \quad \text{for} \quad n \geq 0$$

and so when

$$p > 1$$

the zero equilibrium of Eq(4.19) is globally asymptotically stable. On the other hand when

$$p = 1,$$

$$y_{n+1} < y_{n-1} \quad \text{for} \quad n \geq 0$$

and so every positive solution of Eq(4.19) converges to a period-two solution

$$\ldots, \phi, \psi, \phi, \psi, \ldots .$$

Furthermore we can see that

$$\phi\psi = 0$$

but whether there exists solutions with

$$\phi = \psi = 0$$

remains an **open question**.

Finally when

$$p < 1$$

by invoking the stable manifold theory one can see that there exist nonoscillatory solutions which converge monotonically to the positive equilibrium $1-p$. It follows from the identity

$$y_{n+1} - y_{n-1} = \frac{(1-p) - y_n}{p + y_n} y_{n-1}, \quad n = 0, 1, \ldots$$

that every oscillatory solution is such that the subsequences of even and odd terms converge monotonically one to ∞ and the other to 0.

4.7 The Case $\alpha = \beta = A = 0$: $x_{n+1} = \frac{\gamma x_{n-1}}{Bx_n + Cx_{n-1}}$

This is the $(1,2)$-type Eq(4.9) which by the change of variables

$$x_n = \frac{\gamma}{By_n}$$

reduces to the difference equation

$$y_{n+1} = p + \frac{y_{n-1}}{y_n}, \quad n = 0, 1, \ldots \tag{4.20}$$

where

$$p = \frac{C}{B} \in (0, \infty).$$

This equation was investigated in [7] where it was shown that the following statements are true:

(i) $p \geq 1$ is a necessary and sufficient condition for every solution of Eq(4.20) to be bounded;

(ii) When $p = 1$, every solution of Eq(4.20) converges to a period-two solution;

(iii) When $p > 1$, the equilibrium $\overline{y} = p + 1$ of Eq(4.20) is globally asymptotically stable.

We now proceed to establish the above results.

Clearly the only equilibrium point of Eq(4.20) is $\overline{y} = p + 1$.

The linearized equation of Eq(4.20) about the equilibrium point $\overline{y} = p + 1$ is

$$z_{n+1} + \frac{1}{p+1} z_n - \frac{1}{p+1} z_{n-1} = 0, \quad n = 0, 1, \ldots . \tag{4.21}$$

As a simple consequence of the linearized stability Theorem 1.1.1 we obtain the following result:

Theorem 4.7.1 *The equilibrium point $\overline{y} = p + 1$ of Eq(4.20) is locally asymptotically stable when $p > 1$ and is an unstable saddle point when $p < 1$.*

The next result about semicycles is an immediate consequence of some straightforward arguments and Theorem 1.7.1.

Theorem 4.7.2 *(a) Let $\{y_n\}_{n=-1}^{\infty}$ be a solution of Eq(4.20) which consists of at least two semicycles. Then $\{y_n\}_{n=-1}^{\infty}$ is oscillatory. Moreover, with the possible exception of the first semicycle, every semicycle has length one.*

(b) Let $\{y_n\}_{n=-1}^{\infty}$ be a solution of Eq(4.20) which consists of a single semicycle. Then $\{y_n\}_{n=-1}^{\infty}$ converges monotonically to $\overline{y} = p + 1$.

4.7.1 The Case $p < 1$

Here we show that there exist solutions of Eq(4.20) which are unbounded. In fact the following result is true.

Theorem 4.7.3 *Let $p < 1$, and let $\{y_n\}_{n=-1}^{\infty}$ be a solution of Eq(4.20) such that $0 < y_{-1} \leq 1$ and $y_0 \geq \frac{1}{1-p}$. Then the following statements are true:*

1. $\lim_{n\to\infty} y_{2n} = \infty$.

2. $\lim_{n\to\infty} y_{2n+1} = p$.

Proof. Note that $\frac{1}{1-p} > p + 1$, and so $y_0 > p + 1$. It suffices to show that

$$y_1 \in (p, 1] \quad \text{and} \quad y_2 \geq p + y_0$$

and use induction to complete the proof.

Indeed $y_1 = p + \frac{y_1}{y_0} > p$. Also,

$$y_1 = p + \frac{y_{-1}}{y_0} \leq p + \frac{1}{y_0} \leq 1,$$

and so $y_1 \in (p, 1]$. Hence $y_2 = p + \frac{y_0}{y_1} \geq p + y_0$. □

4.7.2 The Case $p = 1$

The following result follows from Lemma 2.7.1 and some straightforward arguments. See also Section 2.5.

Theorem 4.7.4 *Let $p = 1$, and let $\{y_n\}_{n=-1}^{\infty}$ be a solution of Eq(4.20). Then the following statements are true.*

(a) Suppose $\{y_n\}_{n=-1}^{\infty}$ consists of a single semicycle. Then $\{y_n\}_{n=-1}^{\infty}$ converges monotonically to $\overline{y} = 2$.

(b) Suppose $\{y_n\}_{n=-1}^{\infty}$ consists of at least two semicycles. Then $\{y_n\}_{n=-1}^{\infty}$ converges to a prime period-two solution of Eq(4.20).

(c) Every solution of Eq(4.20) converges to a period-two solution if and only if $p = 1$.

4.7.3 The Case $p > 1$

Here we show that the equilibrium point $\overline{y} = p+1$ of Eq(4.20) is globally asymptotically stable. The following result will be useful.

Lemma 4.7.1 *Let $p > 1$, and let $\{y_n\}_{n=-1}^{\infty}$ be a solution of Eq(4.20). Then*

$$p + \frac{p-1}{p} \leq \liminf_{n\to\infty} y_n \leq \limsup_{n\to\infty} y_n \leq \frac{p^2}{p-1}.$$

Proof. It follows from Theorem 4.7.2 that we may assume that every semicycle of $\{y_n\}_{n=-1}^{\infty}$ has length one, that $p < y_n$ for all $n \geq -1$, and that $p < y_0 < p+1 < y_{-1}$.

We shall first show that $\limsup_{n\to\infty} y_n \leq \frac{p^2}{p-1}$. Note that for $n \geq 0$,

$$y_{2n+1} < p + \frac{y_{2n-1}}{p}.$$

So as every solution of the difference equation

$$y_{m+1} = p + \frac{1}{p} y_m, \quad m = 0, 1, \ldots$$

converges to $\frac{p^2}{p-1}$, it follows that $\limsup_{n\to\infty} y_n \le \frac{p^2}{p-1}$.

We shall next show that $p + \frac{p-1}{p} \le \liminf_{n\to\infty} y_n$. Let $\epsilon > 0$. There clearly exists $N \ge 0$ such that for all $n \ge N$,

$$y_{2n-1} < \frac{p^2+\epsilon}{p-1}.$$

Let $n \ge N$. Then

$$y_{2n} = p + \frac{y_{2n-2}}{y_{2n-1}} > p + p\frac{p-1}{p^2+\epsilon} = \frac{p^3 + p\epsilon + p(p-1)}{p^2+\epsilon}.$$

So as ϵ is arbitrary, we have

$$\liminf_{n\to\infty} y_n \ge \frac{p^3 + p(p-1)}{p^2} = p + \frac{p-1}{p}.$$

□

We are now ready for the following result.

Theorem 4.7.5 *Let $p > 1$. Then the equilibrium $\overline{y} = p + 1$ is globally asymptotically stable.*

Proof. We know by Theorem 4.7.1 that $\overline{y} = p+1$ is a locally asymptotically stable equilibrium point of Eq(4.20). So let $\{y_n\}_{n=-1}^{\infty}$ be a solution of Eq(4.20). It suffices to show that

$$\lim_{n\to\infty} y_n = p + 1.$$

For $x, y \in (0, \infty)$, set

$$f(x,y) = p + \frac{y}{x}.$$

Then $f \in C[(0,\infty)\times(0,\infty),(0,\infty)]$, f is decreasing in $x \in (0,\infty)$ for each $y \in (0,\infty)$, and f is increasing in $y \in (0,\infty)$ for each $x \in (0,\infty)$. Recall that by Theorem 4.7.4, there exist no solutions of Eq(4.20) with prime period two. Let $\epsilon > 0$, and set

$$a = p \quad \text{and} \quad b = \frac{p^2+\epsilon}{p-1}.$$

Note that

$$f\left(\frac{p^2+\epsilon}{p-1}, p\right) = p + p\frac{p-1}{p^2+\epsilon} > p$$

and

$$f\left(p, \frac{p^2+\epsilon}{p-1}\right) = \frac{p^3+\epsilon p}{p^2-p} = \frac{p^2+\epsilon}{p-1}.$$

Hence

$$p < f(x,y) < \frac{p^2+\epsilon}{p-1} \quad \text{for all} \quad x, y \in \left[p, \frac{p^2+\epsilon}{p-1}\right].$$

Finally note that by Lemma 4.7.1,

$$p < p + \frac{p-1}{p} \le \liminf_{n\to\infty} y_n \le \limsup_{n\to\infty} y_n \le \frac{p^2}{p-1} < \frac{p^2+\epsilon}{p-1}$$

and so by Theorem 1.4.6,

$$\lim_{n\to\infty} y_n = p+1.$$

□

4.8 Open Problems and Conjectures

Conjecture 4.8.1 *Consider the difference equation*

$$x_{n+1} = \frac{A}{x_{n-k}} + \frac{B}{x_{n-l}}, \quad n = 0, 1, \ldots \tag{4.22}$$

where

$$A, B \in (0, \infty) \quad \text{and} \quad k, l \in \{0, 1, \ldots\} \quad \text{with } k < l.$$

and where the initial conditions $x_{-l}, \ldots, x_0$ are arbitrary positive real numbers.

Show that every solution of Eq(4.22) converges, either to the equilibrium or to a periodic solution with period $p \ge 2$ and that what happens and the exact value of p are uniquely determined from the characteristic roots of the linearized equation

$$\lambda^{l+1} + \frac{A}{A+B}\lambda^{l-k} + \frac{B}{A+B} = 0.$$

(See [19], [23], [25], [60], and [61].)

Open Problem 4.8.1 *Let*

$$\ldots, \phi_1, \phi_2, \ldots, \phi_p, \ldots$$

be a given prime period p solution of Eq(4.22). Determine the set of all positive initial conditions $x_{-l}, \ldots, x_0$ such that $\{x_n\}_{n=-l}^{\infty}$ converges to this periodic solution.

Open Problem 4.8.2 (see [18]) *Assume that $A, B \in (0, \infty)$.*

(a) Find all initial conditions y_{-1}, y_0 with

$$y_{-1}y_0 < 0$$

for which the equation

$$y_{n+1} = \frac{A}{y_n} + \frac{B}{y_{n-1}}, \tag{4.23}$$

is well defined for all $n \ge 0$.

(b) Assume that the initial conditions y_{-1}, y_0 are such that $y_{-1}y_0 < 0$ and such that Eq(4.23) is well defined for all $n \ge 0$. Investigate the global behavior of the solution $\{y_n\}$. Is it bounded? Does $\lim_{n\to\infty} y_n$ exist? When does $\{y_n\}$ converge to a two cycle?

Conjecture 4.8.2 *Assume that* $p = 1$. *Show that Eq(4.19) has a solution which converges to zero.*

Open Problem 4.8.3 *Assume that* $p < 1$. *Determine the set of initial conditions*

$$y_{-1}, y_0 \in (0, \infty)$$

for which the solution $\{y_n\}_{n=-1}^{\infty}$ *of Eq(4.19) is bounded.*

Open Problem 4.8.4 *Determine the set* G *of all initial conditions* $(y_{-1}, y_0) \in R \times R$ *through which the equation*

$$y_{n+1} = \frac{y_{n-1}}{1 + y_n}$$

is well defined for all $n \geq 0$ *and, for these initial points, investigate the global character of* $\{y_n\}_{n=-1}^{\infty}$.

Conjecture 4.8.3 *Show that Eq(4.20) possesses a solution* $\{y_n\}_{n=-1}^{\infty}$ *which remains above the equilibrium for all* $n \geq -1$.

Open Problem 4.8.5 *(a) Assume* $p \in (0, \infty)$. *Determine all initial conditions* $y_{-1}, y_0 \in \mathbf{R}$ *such that the equation (4.20) is well defined for all* $n \geq 0$.

(b) Let $y_{-1}, y_0 \in \mathbf{R}$ *be such that Eq(4.20) is well defined for all* $n \geq 0$. *For such an initial point, investigate the boundedness, the asymptotic behavior, and the periodic nature of the solution* $\{y_n\}_{n=-1}^{\infty}$ *of Eq(4.20).*

(c) Discuss (a) and (b) when $p < 0$.

Open Problem 4.8.6 (See [13]) *Find the set* $\mathcal{G}$ *of all initial points* $(x_{-1}, x_0) \in R \times R$ *through which the equation*

$$x_{n+1} = -1 + \frac{x_{n-1}}{x_n} \tag{4.24}$$

is well defined for all $n \geq 0$.

Conjecture 4.8.4 *Assume that* $(x_{-1}, x_0) \in R \times R$ *is such that the equation (4.24) is well defined for all* $n \geq 0$ *and bounded. Show that* $\{x_n\}_{n=-1}^{\infty}$ *is a three cycle.*

Open Problem 4.8.7 *Let $f \in C[(0,\infty),(0,\infty)]$. Obtain necessary and sufficient conditions on f for every positive solution of the equation*

$$x_{n+1} = \frac{f(x_n)}{x_n + x_{n-1}}, \quad n = 0, 1, \ldots$$

to be bounded.

Open Problem 4.8.8 *Let $a, a_i \in (0,\infty)$ for $i = 0, \ldots, k$. Investigate the global asymptotic stability of the equilibrium points of the difference equation*

$$x_{n+1} = \frac{x_n}{a + a_0 x_n + \ldots + a_k x_{n-k}}, \quad n = 0, 1, \ldots$$

with positive initial conditions.

Conjecture 4.8.5 (See [27]) *Assume $r \in (0,\infty)$. Show that every positive solution of the* **population model**

$$J_{n+1} = J_{n-1} e^{r-(J_n + J_{n-1})}, \quad n = 0, 1, \ldots$$

converges to a period-two solution.

Conjecture 4.8.6 (**Pielou's Discrete Logistic Model**; see [55], [66], and [67].) *Assume*

$$\alpha \in (1,\infty) \quad \textit{and} \quad k \in \{0, 1, \ldots\}.$$

Show that the positive equilibrium of Pielou's discrete logistic model

$$x_{n+1} = \frac{\alpha x_n}{1 + x_{n-k}}, \quad n = 0, 1, \ldots$$

with positive initial conditions is globally asymptotically stable if and only if

$$\frac{\alpha - 1}{\alpha} < 2\cos\frac{k\pi}{2k+1}.$$

Open Problem 4.8.9 (See [5]) *For which initial values $x_{-1}, x_0 \in (0,\infty)$ does the sequence defined by*

$$x_{n+1} = 1 + \frac{x_{n-1}}{x_n}, \quad n = 0, 1, \ldots$$

converge to two?

(See also Section 4.7)

Open Problem 4.8.10 (See [31]) *Consider the difference equation*

$$y_{n+1} = y_{n-1}e^{-y_n}, \quad n = 0, 1, \ldots \tag{4.25}$$

with nonnegative initial conditions.

One can easily see that every solution $\{y_n\}_{n=-1}^{\infty}$ *of Eq(4.25) converges to a period-two solution*

$$\ldots, \ell, 0, \ell, 0, \ldots .$$

More precisely, each of the subsequences $\{y_{2n}\}_{n=0}^{\infty}$ *and* $\{y_{2n+1}\}_{n=-1}^{\infty}$ *is decreasing and*

$$\left[\lim_{n\to\infty} y_{2n}\right]\left[\lim_{n\to\infty} y_{2n+1}\right] = 0.$$

(a) *Find all initial points* $y_{-1}, y_0 \in [0, \infty)$ *through which the solutions of Eq(4.25) converge to zero. In other words, find the basin of attraction of the equilibrium of Eq(4.25).*

(b) *Determine the limits of the subsequences of even and odd terms of every solution of Eq(4.25), in terms of the initial conditions* y_{-1} *and* y_0 *of the solution.*

Open Problem 4.8.11 *Assume* $p \in (0, 1)$.

(a) *Find the set* B *of all initial conditions* $y_{-1}, y_0 \in (0, \infty)$ *such that the solutions* $\{y_n\}_{n=-1}^{\infty}$ *of Eq(4.20) are bounded.*

(b) *Let* $y_{-1}, y_0 \in B$. *Investigate the asymptotic behavior of* $\{y_n\}_{n=-1}^{\infty}$.

Open Problem 4.8.12 *Let* $\{p_n\}_{n=0}^{\infty}$ *be a convergent sequence of nonnegative real numbers with a finite limit,*

$$p = \lim_{n\to\infty} p_n.$$

Investigate the asymptotic behavior and the periodic nature of all positive solutions of each of the following difference equations:

$$y_{n+1} = \frac{p_n}{y_n} + \frac{1}{y_{n-1}}, \quad n = 0, 1, \ldots \tag{4.26}$$

$$y_{n+1} = \frac{p_n y_n}{1 + y_{n-1}}, \quad n = 0, 1, \ldots \tag{4.27}$$

$$y_{n+1} = \frac{p_n + y_n}{p_n + y_{n-1}}, \quad n = 0, 1, \dots \tag{4.28}$$

$$y_{n+1} = \frac{y_{n-1}}{p_n + y_n}, \quad n = 0, 1, \dots \tag{4.29}$$

$$y_{n+1} = p_n + \frac{y_{n-1}}{y_n}, \quad n = 0, 1, \dots . \tag{4.30}$$

Open Problem 4.8.13 *Let* $\{p_n\}_{n=0}^{\infty}$ *be a periodic sequence of nonnegative real numbers with period* $k \geq 2$. *Investigate the global character of all positive solutions of each of Eqs(4.26) - (4.30).*

Open Problem 4.8.14 *For each of the following difference equations determine the "good" set* $\mathcal{G} \subset \mathcal{R} \times \mathcal{R}$ *of all initial conditions* $(y_{-1}, y_0) \in R \times R$ *through which the equation is well defined for all* $n \geq 0$. *Then for every* $(y_{-1}, y_0) \in \mathcal{G}$, *investigate the long term behavior of the solution* $\{y_n\}_{n=-1}^{\infty}$:

$$y_{n+1} = \frac{y_n}{1 - y_{n-1}}, \tag{4.31}$$

$$y_{n+1} = \frac{1 + y_n}{1 - y_{n-1}}, \tag{4.32}$$

$$y_{n+1} = \frac{y_{n-1}}{1 - y_n}, \tag{4.33}$$

$$y_{n+1} = -1 + \frac{y_{n-1}}{y_n}. \tag{4.34}$$

For those equations from the above list for which you were successful, extend your result by introducing arbitrary real parameters in the equation.

Chapter 5

(2, 1)-Type Equations

5.1 Introduction

Eq(1) contains the following nine equations of the (2, 1)-type:

$$x_{n+1} = \frac{\alpha + \beta x_n}{A}, \quad n = 0, 1, \ldots \tag{5.1}$$

$$x_{n+1} = \frac{\alpha + \beta x_n}{B x_n}, \quad n = 0, 1, \ldots \tag{5.2}$$

$$x_{n+1} = \frac{\alpha + \beta x_n}{C x_{n-1}}, \quad n = 0, 1, \ldots \tag{5.3}$$

$$x_{n+1} = \frac{\alpha + \gamma x_{n-1}}{A}, \quad n = 0, 1, \ldots \tag{5.4}$$

$$x_{n+1} = \frac{\alpha + \gamma x_{n-1}}{B x_n}, \quad n = 0, 1, \ldots \tag{5.5}$$

$$x_{n+1} = \frac{\alpha + \gamma x_{n-1}}{C x_{n-1}}, \quad n = 0, 1, \ldots \tag{5.6}$$

$$x_{n+1} = \frac{\beta x_n + \gamma x_{n-1}}{A}, \quad n = 0, 1, \ldots \tag{5.7}$$

$$x_{n+1} = \frac{\beta x_n + \gamma x_{n-1}}{Bx_n}, \quad n = 0, 1, \ldots \tag{5.8}$$

and

$$x_{n+1} = \frac{\beta x_n + \gamma x_{n-1}}{Cx_{n-1}}, \quad n = 0, 1, \ldots. \tag{5.9}$$

Please recall our classification convention in which all parameters that appear in these equations are positive, the initial conditions are nonnegative, and the denominators are always positive.

Of these nine equations, Eqs(5.1), (5.4) and (5.7) are linear. Eq(5.2) is a Riccati equation and Eq(5.6) is essentially like Eq(5.2).

The change of variables

$$x_n = \frac{\gamma}{B} y_n,$$

reduces Eq(5.8) to Eq(4.20) with $p = \frac{\beta}{\gamma}$, which was investigated in Section 4.7.

Finally the change of variables

$$x_n = \frac{\gamma}{C} + \frac{\beta}{C} y_n,$$

reduces Eq(5.9) to Eq(4.18) with $p = \frac{\gamma}{\beta}$, which was investigated in Section 4.5.

Therefore there remain Eqs(5.3) and (5.5), which will be investigated in the next two sections.

5.2 The Case $\gamma = A = B = 0$: $x_{n+1} = \frac{\alpha + \beta x_n}{Cx_{n-1}}$ - Lyness' Equation

This is the (2, 1)-type Eq(5.3) which by the change of variables reduces to the equation

$$y_{n+1} = \frac{p + y_n}{y_{n-1}}, \quad n = 0, 1, \ldots \tag{5.10}$$

where

$$p = \frac{\alpha C}{\beta^2}.$$

The special case of Eq(5.10) where

$$p = 1$$

was discovered by Lyness in 1942 while he was working on a problem in Number Theory. (See ([42], p. 131) for the history of the problem.) In this special case, the equation becomes

$$y_{n+1} = \frac{1+y_n}{y_{n-1}}, \quad n = 0, 1, \ldots \tag{5.11}$$

every solution of which is periodic with period five. Indeed the solution of Eq(5.11) with initial conditions y_{-1} and y_0 is the five-cycle:

$$y_{-1}, y_0, \frac{1+y_0}{y_{-1}}, \frac{1+y_{-1}+y_0}{y_{-1}y_0}, \frac{1+y_{-1}}{y_0}, \ldots .$$

Eq(5.10) possesses the **invariant**

$$I_n = (p + y_{n-1} + y_n)\left(1 + \frac{1}{y_{n-1}}\right)\left(1 + \frac{1}{y_n}\right) = constant \tag{5.12}$$

from which it follows that every solution of Eq(5.10) is bounded from above and from below by positive constants.

It was shown in [30] that no nontrivial solution of Eq(5.10) has a limit.

It was also shown in [44] by using KAM theory that the positive equilibrium $\overline{y}$ of Eq(5.10) is stable but not asymptotically stable. This result was also established in [49] by using a Lyapunov function.

Concerning the semicycles of solutions of Eq(5.10) the following result was established in [43]:

Theorem 5.2.1 *Assume that $p > 0$. Then the following statements about the positive solutions of Eq(5.10) are true:*

(a) The absolute extreme in a semicycle occurs in the first or in the second term.

(b) Every nontrivial semicycle, after the first one, contains at least two and at most three terms.

There is substantial literature on Lyness' Equation. See [11], [42], [43], [57]-[59], [71], and [72] and the references cited therein.

5.3 The Case $\beta = A = C = 0$: $x_{n+1} = \frac{\alpha + \gamma x_{n-1}}{Bx_n}$

This is the (2, 1)-type Eq(5.5) which by the change of variables

$$x_n = \frac{\gamma}{B} y_n$$

reduces to the equation

$$y_{n+1} = \frac{p + y_{n-1}}{y_n}, \quad n = 0, 1, \ldots \tag{5.13}$$

where

$$p = \frac{\alpha B}{\gamma^2}.$$

Eq(5.13) was investigated in [28].

By the linearized stability Theorem 1.1.1 one can see that the unique equilibrium point

$$\overline{y} = \frac{1 + \sqrt{1 + 4p}}{2}$$

of Eq(5.13) is a saddle point.

By Theorem 1.7.1 one can see that except possibly for the first semicycle, every oscillatory solution of Eq(5.13) has semicycles of length one.

Concerning the nonoscillatory solutions of Eq(5.13) one can easily see that every nonoscillatory solution of Eq(5.13) converges monotonically to the equilibrium $\overline{y}$.

The global character of solutions of Eq(5.13) is a consequence of the identity

$$y_{n+1} - y_{n-1} = \frac{y_{n-1} - y_{n-2}}{y_n} \quad \text{for} \quad n \geq 1.$$

From this it follows that a solution of Eq(5.13) either approaches the positive equilibrium $\overline{y}$ monotonically or it oscillates about $\overline{y}$ with semicycles of length one in such a way that $\{y_{2n}\}$ and $\{y_{2n+1}\}$ converge monotonically one to zero and other to ∞.

5.4 Open Problems and Conjectures

By using the invariant (5.12), one can easily see that every solution of Lyness' Eq(5.10) is bounded from above and from below by positive numbers. To this day we have not found a proof for the boundedness of solutions of Eq(5.10) without using essentially the invariant.

Open Problem 5.4.1 *Show that every solution of Lyness' Eq(5.10) is bounded without using the invariant (5.12).*

Open Problem 5.4.2 *Assume that $f \in C[(0,\infty),(0,\infty)]$. Obtain necessary and sufficient conditions in terms of f so that every positive solution of the difference equation*

$$x_{n+1} = \frac{f(x_n)}{x_{n-1}}, \quad n = 0, 1, \ldots \tag{5.14}$$

is bounded.

What is it that makes Lyness' equation possess an invariant? What type of difference equations possess an invariant? Along these lines, we offer the following open problems and conjectures.

Open Problem 5.4.3 *Assume that $f \in C[(0,\infty),(0,\infty)]$. Obtain necessary and sufficient conditions in terms of f so that the difference equation (5.14) possesses a (nontrivial) invariant.*

Conjecture 5.4.1 *Assume that $f \in C[(0,\infty),(0,\infty)]$ and that the difference equation (5.14) possesses a unique positive equilibrium point $\overline{x}$ and a nontrivial invariant. Show that the linearized equation of Eq(5.14) about $\overline{x}$ does not have both eigenvalues in $|\lambda| < 1$ and does not have both eigenvalues in $|\lambda| > 1$.*

When we allow the initial conditions to be real numbers the equation

$$y_{n+1} = \frac{p + y_n}{y_{n-1}} \tag{5.15}$$

may not even be well defined for all $n \geq 0$. To illustrate, we offer the following open problems and conjectures.

Open Problem 5.4.4 (See [26].) *Assume $p \in (-\infty,\infty)$. Determine the set $\mathcal{G}$ of all initial conditions $y_{-1}, y_0 \in (-\infty,\infty)$ through which Eq(5.15) is well defined for all $n \geq 0$.*

Conjecture 5.4.2 *Let $\mathcal{G}$ be the set of initial conditions defined in the Open problem 5.4.4. Show that the subset of $\mathcal{G}$ through which the solutions of Eq(5.15) are unbounded is a set of measure zero.*

Conjecture 5.4.3 *Assume $p \in (-\infty,\infty)$. Show that the set of points $(y_{-1}, y_0) \in (-\infty,\infty) \times (-\infty,\infty)$ through which Eq(5.15) is not well defined for all $n \geq 0$, is a set of points without interior.*

Open Problem 5.4.5 *Determine all positive integers k with the property that every positive solution of the equation*

$$y_{n+1} = \frac{1 + \ldots + y_{n-k}}{y_{n-k-1}}, \quad n = 0, 1, \ldots$$

is periodic.

Conjecture 5.4.4 *Show that every positive solution of the equation*

$$y_{n+1} = \frac{y_n}{y_{n-1}} + \frac{y_{n-4}}{y_{n-3}}, \quad n = 0, 1, \ldots$$

converges to a period-six solution.

Conjecture 5.4.5 (See [20]) *Show that every positive solution of the equation*

$$y_{n+1} = \frac{1}{y_n y_{n-1}} + \frac{1}{y_{n-3} y_{n-4}}, \quad n = 0, 1, \ldots$$

converges to a period-three solution.

Conjecture 5.4.6 *Show that the equation*

$$y_{n+1} = \frac{1 + y_{n-1}}{y_n}, \quad n = 0, 1, \ldots$$

has a nontrivial positive solution which decreases monotonically to the equilibrium of the equation.

Open Problem 5.4.6 *Find the set $\mathcal{G}$ of all initial points $(y_{-1}, y_0) \in R \times R$ through which the equation*

$$y_{n+1} = \frac{1 + y_{n-1}}{y_n}$$

is well defined for all $n \geq 0$. Determine the periodic character and the asymptotic behavior of all solutions with $(y_{-1}, y_0) \in \mathcal{G}$.

Conjecture 5.4.7 (**May's Host Parasitoid Model**; see [59]) *Assume* $\alpha > 1$. *Show that every positive solution of May's Host Parasitoid Model*

$$x_{n+1} = \frac{\alpha x_n^2}{(1+x_n)x_{n-1}}, \quad n = 0, 1, \ldots$$

is bounded.

Conjecture 5.4.8 (**The Gingerbreadman Map**; see [59]) *Let* $A \in (0, \infty)$. *Show that every positive solution of the equation*

$$x_{n+1} = \frac{\max\{x_n^2, A\}}{x_n x_{n-1}}, \quad n = 0, 1, \ldots \tag{5.16}$$

is bounded.

Note that the change of variables

$$x_n = \begin{cases} A^{\frac{1+y_n}{2}} & \text{if } A > 1 \\ e^{\frac{y_n}{2}} & \text{if } A = 1 \\ A^{\frac{-1+y_n}{2}} & \text{if } A < 1 \end{cases}$$

reduces Eq(5.16) to

$$y_{n+1} = |y_n| - y_{n-1} + \delta, \quad n = 0, 1, \ldots \tag{5.17}$$

where

$$\delta = \begin{cases} 1 & \text{if } A < 1 \\ 0 & \text{if } A = 1 \\ -1 & \text{if } A > 1. \end{cases}$$

When $\delta = 1$, Eq(5.17) is called the **Gingerbreadman Map** and was investigated by Devaney [17].

Chapter 6

(2, 2)-Type Equations

6.1 Introduction

Eq(1) contains the following nine equations of the (2, 2)-type:

$$x_{n+1} = \frac{\alpha + \beta x_n}{A + Bx_n}, \quad n = 0, 1, \ldots \tag{6.1}$$

$$x_{n+1} = \frac{\alpha + \beta x_n}{A + Cx_{n-1}}, \quad n = 0, 1, \ldots \tag{6.2}$$

$$x_{n+1} = \frac{\alpha + \beta x_n}{Bx_n + Cx_{n-1}}, \quad n = 0, 1, \ldots \tag{6.3}$$

$$x_{n+1} = \frac{\alpha + \gamma x_{n-1}}{A + Bx_n}, \quad n = 0, 1, \ldots \tag{6.4}$$

$$x_{n+1} = \frac{\alpha + \gamma x_{n-1}}{A + Cx_{n-1}}, \quad n = 0, 1, \ldots \tag{6.5}$$

$$x_{n+1} = \frac{\alpha + \gamma x_{n-1}}{Bx_n + Cx_{n-1}}, \quad n = 0, 1, \ldots \tag{6.6}$$

$$x_{n+1} = \frac{\beta x_n + \gamma x_{n-1}}{A + Bx_n}, \quad n = 0, 1, \dots \tag{6.7}$$

$$x_{n+1} = \frac{\beta x_n + \gamma x_{n-1}}{A + Cx_{n-1}}, \quad n = 0, 1, \dots \tag{6.8}$$

$$x_{n+1} = \frac{\beta x_n + \gamma x_{n-1}}{Bx_n + Cx_{n-1}}, \quad n = 0, 1, \dots . \tag{6.9}$$

Please recall our classification convention in which all parameters that appear in these equations are positive, the initial conditions are nonnegative, and the denominators are always positive.

Eq(6.5) is essentially similar to Eq(6.1). In fact if $\{x_n\}_{n=-1}^{\infty}$ is a positive solution of Eq(6.5) then $\{x_{2n}\}_{n=0}^{\infty}$ and $\{x_{2n-1}\}_{n=0}^{\infty}$ are both solutions of the Riccati equations

$$z_{n+1} = \frac{\alpha + \gamma z_n}{A + Cz_n}, \quad n = 0, 1, \dots$$

with

$$z_n = x_{2n} \quad \text{for } n \geq 0$$

and

$$z_{n+1} = \frac{\alpha + \gamma z_n}{A + Cz_n}, \quad n = -1, 0, \dots$$

with

$$z_n = x_{2n+1} \quad \text{for } n \geq -1,$$

respectively.

Therefore we will omit any further discussion of Eq(6.5).

6.2 The Case $\gamma = C = 0$: $x_{n+1} = \frac{\alpha+\beta x_n}{A+Bx_n}$ - Riccati Equation

This is the (2,2)-type Eq(6.1) which is in fact a Riccati equation. See Section 1.6.

To avoid a degenerate situation we will assume that

$$\alpha B - \beta A \neq 0.$$

The Riccati number associated with this equation is

$$R = \frac{\beta A - \alpha B}{(\beta + A)^2}$$

and clearly

$$R < \frac{1}{4}.$$

Now the following result is a consequence of Theorem 1.6.2:

Theorem 6.2.1 *The positive equilibrium of Eq(6.1) is globally asymptotically stable.*

6.3 The Case $\gamma = B = 0:\ x_{n+1} = \frac{\alpha+\beta x_n}{A+Cx_{n-1}}$

This is the $(2,2)$-type Eq(6.2) which by the change of variables

$$x_n = \frac{A}{C} y_n$$

reduces to the equation

$$y_{n+1} = \frac{p + qy_n}{1 + y_{n-1}}, \quad n = 0, 1, \ldots \tag{6.10}$$

where

$$p = \frac{\alpha C}{A^2} \quad \text{and} \quad q = \frac{\beta}{A}.$$

(Eq(6.10) was investigated in [42] and [43]. See also [36].)

Eq(6.10) has the unique positive equilibrium $\overline{y}$ given by

$$\overline{y} = \frac{q - 1 + \sqrt{(q-1)^2 + 4p}}{2}.$$

By applying the linearized stability Theorem 1.1.1 we obtain the following result.

Lemma 6.3.1 *The equilibrium $\overline{y}$ of Eq(6.10) is locally asymptotically stable for all values of the parameters p and q.*

Eq(6.10) is a very simple looking equation for which it has long been conjectured that its equilibrium is globally asymptotically stable. To this day, the conjecture has not been proven or refuted.

The main results known about Eq(6.10) are the following:

Theorem 6.3.1 *Every solution of Eq(6.10) is bounded from above and from below by positive constants.*

Proof. Let $\{y_n\}$ be a solution of Eq(6.10). Clearly, if the solution is bounded from above by a constant M, then

$$y_{n+1} \geq \frac{p}{1+M}$$

and so it is also bounded from below. Now assume for the sake of contradiction that the solution is not bounded from above. Then there exists a subsequence $\{y_{1+n_k}\}_{k=0}^{\infty}$ such that

$$\lim_{k\to\infty} n_k = \infty, \quad \lim_{k\to\infty} y_{1+n_k} = \infty, \quad \text{and} \quad y_{1+n_k} = \max\{y_n : n \leq n_k\} \quad \text{for } k \geq 0.$$

From (6.10) we see that

$$y_{n+1} < qy_n + p \quad \text{for} \quad n \geq 0$$

and so

$$\lim_{k\to\infty} y_{n_k} = \lim_{k\to\infty} y_{n_k-1} = \infty.$$

Hence, for sufficiently large k,

$$0 \leq y_{1+n_k} - y_{n_k} = \frac{p + [(q-1) - y_{n_k-1}]y_{n_k}}{1 + y_{n_k-1}} < 0$$

which is a contradiction and the proof is complete. □

The oscillatory character of solutions for Eq(6.10) is described by the following result:

Theorem 6.3.2 *Let $\{y_n\}_{n=-1}^{\infty}$ be a nontrivial solution of Eq(6.10). Then the following statements are true:*

(i) Every semicycle, except perhaps for the first one, has at least two terms.

(ii) The extreme in each semicycle occurs at either the first term or the second. Furthermore after the first, the remaining terms in a positive semicycle are strictly decreasing and in a negative semicycle are strictly increasing.

(iii) In any two consecutive semicycles, their extrema cannot be consecutive terms.

(iv) Assume

$$q \leq p.$$

Then, except possibly for the first semicycle, every semicycle contains two or three terms. Furthermore, in every semicycle, there is at most one term which follows the extreme.

Proof. We present the proofs for positive semicycles only. The proofs for negative semicycles are similar and will be omitted.

(i) Assume that for some $N \geq 0$,

$$y_{N-1} < \bar{y} \quad \text{and} \quad y_N \geq \bar{y}.$$

Then

$$y_{N+1} = \frac{p + qy_N}{1 + y_{N-1}} > \frac{p + q\bar{y}}{1 + \bar{y}} = \bar{y}.$$

(ii) Assume that for some $N \geq 0$, the first two terms in a positive semicycle are y_N and y_{N+1}. Then

$$y_N \geq \bar{y}, \quad y_{N+1} > \bar{y}$$

and

$$\frac{y_{N+2}}{y_{N+1}} = \frac{1}{y_{N+1}} \frac{p + qy_{N+1}}{1 + y_N} = \frac{\frac{p}{y_{N+1}} + q}{1 + y_N} < \frac{\frac{p}{\bar{y}} + q}{1 + \bar{y}} = 1.$$

(iii) We consider the case where a negative semicycle is followed by a positive one. The opposite case is similar and will be omitted. So assume that for some $N \geq 0$ the last two terms in a negative semicycle are y_{N-1} and y_N with

$$y_{N-1} \geq y_N.$$

Then

$$y_{N+1} = \frac{p + qy_N}{1 + y_{N-1}} < \frac{p + qy_{N+1}}{1 + y_N} = y_{N+2}.$$

(iv) Assume that for some $N \geq 0$, the terms y_N, y_{N+1}, and y_{N+2} are all in a positive semicycle. Then

$$y_{N+3} = \frac{p + qy_{N+2}}{1 + y_{N+1}} < \frac{p + qy_{N+1}}{1 + y_{N+1}} < \frac{p + q\bar{y}}{1 + \bar{y}} = \bar{y}.$$

□

Theorem 6.3.3 *The positive equilibrium of Eq(6.10) is globally asymptotically stable if one of the following two conditions holds:*

(i)

$$q < 1;$$

(ii)

$$q \geq 1$$

and

$$\text{either} \quad p \leq q \quad \text{or} \quad q < p \leq 2(q + 1).$$

Proof.

(i) In view of Lemma 6.3.1, it remains to show that every solution $\{y_n\}_{n=-1}^{\infty}$ of Eq(6.10) tends to $\overline{y}$ as $n \to \infty$. To this end, let

$$i = \liminf_{n\to\infty} y_n \quad \text{and} \quad s = \limsup_{n\to\infty} y_n$$

which by Theorem 6.3.1 exist and are positive numbers. Then Eq(6.10) yields

$$i \geq \frac{p+qi}{1+s} \quad \text{and} \quad s \leq \frac{p+qs}{1+i}.$$

Hence

$$p+(q-1)i \leq is \leq p+(q-1)s$$

and so because $q < 1$,

$$i \geq s$$

from which the result follows.

(ii) The proof when

$$p < q$$

follows by applying Theorem 1.4.2. This result was first established in [43].

For the proof when

$$1 \leq q \leq p \leq 2(q+1)$$

see ([42], Theorem 3.4.3 (f), p. 71).

□

6.4 The Case $\gamma = A = 0$: $x_{n+1} = \frac{\alpha+\beta x_n}{Bx_n+Cx_{n-1}}$

This is the $(2,2)$-type Eq(6.3) which by the change of variables

$$x_n = \frac{\beta}{B} y_n$$

reduces to the difference equation

$$y_{n+1} = \frac{y_n + p}{y_n + q y_{n-1}}, \quad n = 0, 1, \ldots \tag{6.11}$$

where

$$p = \frac{\alpha B}{\beta^2} \quad \text{and} \quad q = \frac{C}{B}.$$

Eq(6.11) was investigated in [15].

The following result is a consequence of the linearized stability Theorem 1.1.1.

Theorem 6.4.1 *The unique positive equilibrium point*

$$\overline{y} = \frac{1+\sqrt{1+4p(q+1)}}{2(q+1)}$$

of Eq(6.11) is locally asymptotically stable for all values of the parameters p and q.

6.4.1 Invariant Intervals

The following result establishes that the interval I with end points 1 and $\frac{p}{q}$ is an **invariant interval** for Eq(6.11) in the sense that if, for some $k \geq 0$, two consecutive values y_{k-1} and y_k of a solution lie in I, then $y_m \in I$ for all $m > k$.

Lemma 6.4.1 *Let $\{y_n\}_{n=-1}^{\infty}$ be a solution of Eq(6.11). Then the following statements are true:*

(a) Assume $p < q$ and suppose that for some $k \geq 0$,

$$y_{k-1}, y_k \in \left[\frac{p}{q}, 1\right].$$

Then

$$y_n \in \left[\frac{p}{q}, 1\right] \quad \textit{for all } n > k.$$

(b) Assume $p > q$ and suppose that for some $k \geq 0$,

$$y_{k-1}, y_k \in \left[1, \frac{p}{q}\right].$$

Then

$$y_n \in \left[1, \frac{p}{q}\right] \quad \textit{for all } n > k.$$

Proof.

(a) The basic ingredient behind the proof is the fact that when $u, v \in [\frac{p}{q}, \infty)$, the function

$$f(u,v) = \frac{u+p}{u+qv}$$

is increasing in u and decreasing in v. Indeed,

$$y_{k+1} = \frac{y_k + p}{y_k + qy_{k-1}} \leq \frac{y_k + p}{y_k + q\frac{p}{q}} = 1,$$

$$y_{k+1} = \frac{y_k + p}{y_k + qy_{k-1}} \geq \frac{\frac{p}{q} + p}{\frac{p}{q} + q} > \frac{p}{q}$$

and the proof follows by induction.

(b) The proof is similar and will be omitted. □

6.4.2 Semicycle Analysis

Let $\{y_n\}_{n=-1}^{\infty}$ be a solution of Eq(6.11). Then observe that the following identities are true:

$$y_{n+1} - 1 = q\frac{\frac{p}{q} - y_{n-1}}{y_n + qy_{n-1}} \quad \text{for } n \geq 0, \tag{6.12}$$

$$y_{n+1} - \frac{p}{q} = \frac{(q-p) + pq(1 - y_{n-1})}{q(y_n + qy_{n-1})} \quad \text{for } n \geq 0, \tag{6.13}$$

and

$$y_n - y_{n+4} = \frac{(y_n - 1)[qy_ny_{n+3} + y_{n+1}y_{n+3}] + qy_{n+1}(y_n - \frac{p}{q})}{q(p + y_{n+1}) + y_{n+3}(y_{n+1} + qy_n)} \quad \text{for } n \geq 0. \tag{6.14}$$

First we will analyze the semicycles of the solution $\{y_n\}_{n=-1}^{\infty}$ under the assumption that

$$p > q. \tag{6.15}$$

The following result is a direct consequence of (6.12)-(6.15):

Lemma 6.4.2 *Assume that (6.15) holds and let $\{y_n\}_{n=-1}^{\infty}$ be a solution of Eq(6.11). Then the following statements are true:*

(i) If for some $N \geq 0$, $y_{N-1} < \frac{p}{q}$, then $y_{N+1} > 1$;

(ii) If for some $N \geq 0$, $y_{N-1} = \frac{p}{q}$, then $y_{N+1} = 1$;

(iii) If for some $N \geq 0$, $y_{N-1} > \frac{p}{q}$, then $y_{N+1} < 1$;

(iv) If for some $N \geq 0$, $y_{N-1} \geq 1$, then $y_{N+1} < \frac{p}{q}$;

(v) If for some $N \geq 0$, $y_{N-1} \leq 1$, then $y_{N+1} > 1$;

(vi) If for some $N \geq 0$, $y_N \geq \frac{p}{q}$, then $y_{N+4} < y_N$;

(vii) If for some $N \geq 0$, $y_N \leq 1$, then $y_{N+4} > y_N$;

(viii) If for some $N \geq 0$, $1 \leq y_{N-1} \leq \frac{p}{q}$, then $1 \leq y_{N+1} \leq \frac{p}{q}$;

(ix) If for some $N \geq 0$, $1 \leq y_{N-1}, y_N \leq \frac{p}{q}$, then $y_n \in [1, \frac{p}{q}]$ for $n \geq N$. That is $[1, \frac{p}{q}]$ is an invariant interval for Eq(6.11);

(x)

$$1 < \overline{y} < \frac{p}{q}.$$

The next result, which is a consequence of Theorem 1.7.2, states that when (6.15) holds, every nontrivial and oscillatory solution of Eq(6.11), which lies in the interval $[1, \frac{p}{q}]$, oscillates about the equilibrium $\overline{y}$ with semicycles of length one or two.

Theorem 6.4.2 *Assume that Eq(6.11) holds. Then every nontrivial and oscillatory solution of Eq(6.11) which lies in the interval $[1, \frac{p}{q}]$, oscillates about $\overline{y}$ with semicycles of length one or two.*

Next we will analyze the semicycles of the solutions $\{y_n\}_{n=-1}^{\infty}$ under the assumption that

$$q = p. \tag{6.16}$$

In this case, Eq(6.11) reduces to

$$y_{n+1} = \frac{y_n + p}{y_n + py_{n-1}}, \qquad n = 0, 1, \ldots \tag{6.17}$$

with the unique equilibrium point

$$\overline{y} = 1.$$

Also identities (6.12)-(6.14) reduce to

$$y_{n+1} - 1 = p\frac{1 - y_{n-1}}{y_n + py_{n-1}} \quad \text{for } n \geq 0 \tag{6.18}$$

and

$$y_n - y_{n+4} = (y_n - 1)\frac{py_{n+1} + py_ny_{n+3} + y_{n+1}y_{n+3}}{p(p + y_{n+1}) + y_{n+3}(y_{n+1} + py_n)} \quad \text{for} \quad n \geq 0. \tag{6.19}$$

and so the following results follow immediately:

Lemma 6.4.3 *Let $\{y_n\}_{n=-1}^{\infty}$ be a solution of Eq(6.17). Then the following statements are true:*

(i) If for some $N \geq 0$, $y_{N-1} < 1$, then $y_{N+1} > 1$;

(ii) If for some $N \geq 0$, $y_{N-1} = 1$, then $y_{N+1} = 1$;

(iii) If for some $N \geq 0$, $y_{N-1} > 1$, then $y_{N+1} < 1$;

(iv) If for some $N \geq 0$, $y_N < 1$, then $y_N < y_{N+4} < 1$;

(v) If for some $N \geq 0$, $y_N > 1$, then $y_N > y_{N+4} > 1$.

Corollary 6.4.1 *Let $\{y_n\}_{n=-1}^{\infty}$ be a nontrivial solution of Eq(6.17). Then $\{y_n\}_{n=-1}^{\infty}$ oscillates about the equilibrium 1 and, except possibly for the first semicycle, the following are true:*

(i) *If $y_{-1} = 1$ or $y_0 = 1$, then the positive semicycle has length three and the negative semicycle has length one.*

(ii) *If $(1 - y_{-1})(1 - y_0) \neq 0$, then every semicycle has length two.*

Finally we will analyze the semicycles of the solutions $\{y_n\}_{n=-1}^{\infty}$ under the assumption that

$$p < q. \tag{6.20}$$

The following result is a direct consequence of (6.12)-(6.14) and (6.20).

Lemma 6.4.4 *Assume that (6.20) holds and let $\{y_n\}_{n=-1}^{\infty}$ be a solution of Eq(6.11). Then the following statements are true:*

(i) *If for some $N \geq 0$, $y_{N-1} < \frac{p}{q}$, then $y_{N+1} > 1$;*

(ii) *If for some $N \geq 0$, $y_{N-1} = \frac{p}{q}$, then $y_{N+1} = 1$;*

(iii) *If for some $N \geq 0$, $y_{N-1} > \frac{p}{q}$, then $y_{N+1} < 1$;*

(iv) *If for some $N \geq 0$, $y_{N-1} \leq 1$, then $y_{N+1} > \frac{p}{q}$;*

(v) *If for some $N \geq 0$, $y_{N-1} \leq \frac{p}{q}$, then $y_{N+1} > \frac{p}{q}$;*

(vi) *If for some $N \geq 0$, $y_N \geq 1$, then $y_{N+4} < y_N$;*

(vii) *If for some $N \geq 0$, $y_N \leq \frac{p}{q}$, then $y_{N+4} > y_N$;*

(viii) *If for some $N \geq 0$, $\frac{p}{q} \leq y_{N-1} \leq 1$, then $\frac{p}{q} \leq y_{N+1} \leq 1$;*

(ix) *If for some $N \geq 0$, $\frac{p}{q} \leq y_{N-1}, y_N \leq 1$, then $y_n \in [\frac{p}{q}, 1]$ for $n \geq N$. That is, $[\frac{p}{q}, 1]$ is an invariant interval for Eq(6.11);*

(x)

$$\frac{p}{q} < \overline{y} < 1.$$

The next result, which is a consequence of Theorem 1.7.4, states that when (6.20) holds, every nontrivial and oscillatory solution of Eq(6.11) which lies in the interval $[\frac{p}{q}, 1]$, after the first semicycle, oscillates with semicycles of length at least two.

Theorem 6.4.3 *Assume that (6.20) holds. Then every nontrivial and oscillatory solution of Eq(6.11) which lies in the interval $[\frac{p}{q}, 1]$, after the first semicycle, oscillates with semicycles of length at least two.*

How do solutions which do not eventually lie in the invariant interval behave?

First assume that

$$p > q$$

and let $\{y_n\}_{n=-1}^{\infty}$ be a solution which does not eventually lie in the interval $I = [1, \frac{p}{q}]$. Then one can see that the solution oscillates relative to the interval $[1, \frac{p}{q}]$, essentially in one of the following two ways:

(i) Two consecutive terms in $(\frac{p}{q}, \infty)$ are followed by two consecutive terms in $(0, 1)$, are followed by two consecutive terms in $(\frac{p}{q}, \infty)$, and so on. The solution never visits the interval $(1, \frac{p}{q})$.

(ii) There exists exactly one term in $(\frac{p}{q}, \infty)$, which is followed by exactly one term in $(1, \frac{p}{q})$, which is followed by exactly one term in $(0, 1)$, which is followed by exactly one term in $(1, \frac{p}{q})$, which is followed by exactly one term in $(\frac{p}{q}, \infty)$, and so on. The solution visits consecutively the intervals

$$\ldots, \left(\frac{p}{q}, \infty\right), \left(1, \frac{p}{q}\right), (0, 1), \left(1, \frac{p}{q}\right), \left(\frac{p}{q}, \infty\right), \ldots$$

in this order with one term per interval.

The situation is essentially the same relative to the interval $[\frac{p}{q}, 1]$ when

$$p < q.$$

6.4.3 Global Stability and Boundedness

Our goal in this section is to establish the following result:

Theorem 6.4.4 *(a) The equilibrium $\overline{y}$ of Eq(6.11) is globally asymptotically stable when*

$$q \leq 1 + 4p. \tag{6.21}$$

(b) Assume that

$$q > 1 + 4p.$$

Then every solution of Eq(6.11) lies eventually in the interval $\left[\frac{p}{q}, 1\right]$.

Proof.

(a) We have already shown that the equilibrium $\overline{y}$ is locally asymptotically stable so it remains to be shown that $\overline{y}$ is a global attractor of every solution $\{y_n\}_{n=-1}^{\infty}$ of Eq(6.11).

Case 1 Assume $p = q$. It follows from (6.19) that each of the four subsequences

$$\{y_{4n+i}\}_{n=0}^{\infty} \quad \text{for} \quad i = 1, 2, 3, 4$$

is either identically equal to one or else it is strictly monotonically convergent. Set

$$L_i = \lim_{n\to\infty} y_{4n+i} \quad \text{for} \quad i = 1, 2, 3, 4.$$

Thus clearly

$$\ldots, L_1, L_2, L_3, L_4, \ldots \tag{6.22}$$

is a periodic solution of Eq(6.11) with period four. By applying (6.19) to the solution (6.22) we see that

$$L_i = 1 \quad \text{for} \quad i = 1, 2, 3, 4$$

and so

$$\lim_{n\to\infty} y_n = 1.$$

Case 2 Assume $p \neq q$. First, we assume that $p > q$.

Recall from Lemma 6.4.2 that $[1, \frac{p}{q}]$ is an invariant interval. In this interval the function

$$f(u, v) = \frac{u + p}{u + qv}$$

is decreasing in both variables and so it follows by applying Theorem 1.4.7 that every solution of Eq(6.11) with two consecutive values in $[1, \frac{p}{q}]$ converges to $\overline{y}$. By using an argument similar to that in the case $p = q$, we can see that if a solution is not eventually in $[1, \frac{p}{q}]$, it converges to a periodic solution with period-four which is not the equilibrium. By applying (6.14) to this period four solution we obtain a contradiction and so every solution of Eq(6.11) must lie eventually in $[1, \frac{p}{q}]$.

Hence for every solution $\{y_n\}_{n=-1}^{\infty}$ of Eq(6.11)

$$\lim_{n\to\infty} y_n = \overline{y}$$

and the proof is complete.

Now, assume that $p < q$.

As in Case 2 we can see that every solution of Eq(6.11) must eventually lie in the interval $[\frac{p}{q}, 1]$. In this interval the function

$$f(u, v) = \frac{u + p}{u + qv}$$

is increasing in u and decreasing in v. Under the assumption that $q \leq 1 + 4p$, Theorem 1.4.5 applies and so every solution of Eq(6.11) converges to $\overline{y}$.

(b) The proof follows from the discussion in part *(a)*.

□

6.5 The Case $\beta = C = 0: \; x_{n+1} = \frac{\alpha+\gamma x_{n-1}}{A+Bx_n}$

This is the $(2,2)$-type Eq(6.4) which by the change of variables

$$x_n = \frac{A}{B} y_n$$

reduces to the difference equation

$$y_{n+1} = \frac{p + q y_{n-1}}{1 + y_n}, \quad n = 0, 1, \ldots \tag{6.23}$$

where

$$p = \frac{\alpha B}{A^2} \quad \text{and} \quad q = \frac{\gamma}{A}.$$

(Eq(6.23) was investigated in [28].)

The only equilibrium point of Eq(6.23) is

$$\overline{y} = \frac{q - 1 + \sqrt{(q-1)^2 + 4p}}{2}$$

and by applying the linearized stability Theorem 1.1.1 we obtain the following result:

Theorem 6.5.1 *The equilibrium $\overline{y}$ of Eq(6.23) is locally asymptotically stable when*

$$q < 1,$$

and is an unstable saddle point when

$$q > 1.$$

Concerning prime period-two solutions, the following result is a consequence of the results in Section 2.7.

Theorem 6.5.2 *(a) Eq(6.23) has a prime period-two solution*

$$\ldots, \phi, \psi, \phi, \psi, \ldots \tag{6.24}$$

if and only if

$$q = 1. \tag{6.25}$$

(b) Assume (6.25) holds. Then the values ϕ and ψ of all prime period-two solutions (6.24) are given by

$$\{\phi, \psi \in (0, \infty) : \phi\psi = p\}.$$

(c) Assume that (6.25) holds. Then every solution of Eq(6.23) converges to a period two solution.

6.5.1 Semicycle Analysis

Our results on the semicycles of solutions for Eq(6.23) are as follows:

Theorem 6.5.3 *Let $\{y_n\}$ be a nontrivial solution of Eq(6.23) and let $\overline{y}$ denote the unique positive equilibrium of Eq(6.23). Then the following statements are true:*

(a) After the first semicycle, an oscillatory solution $\{y_n\}$ of Eq(6.23) oscillates about the equilibrium $\overline{y}$ with semicycles of length one.

(b) Assume $p \geq q$. Then every nonoscillatory solution of Eq(6.23) converges monotonically to the equilibrium $\overline{y}$.

Proof. *(a)* This follows immediately from Theorem 1.7.1

(b) Let $\{y_n\}$ be a nonoscillatory solution of Eq(6.23). We will assume that there exists a positive integer K such that $y_{n-1} \geq \overline{y}$ for every $n \geq K$. The case where $y_{n-1} < \overline{y}$ for $n \geq K$ is similar and will be omitted. It is sufficient to show that $\{y_n\}$ is a decreasing sequence for $n \geq K$. So assume for the sake of contradiction that for some $n_0 \geq K$,

$$y_{n_0} > y_{n_0-1}.$$

Clearly, the condition $p \geq q$ implies that the function

$$f(x) = \frac{p + qx}{1 + x}$$

is decreasing. Then

$$y_{n_0+1} = \frac{p + qy_{n_0-1}}{1 + y_{n_0}} = f(y_{n_0}, y_{n_0-1}) < f(y_{n_0}, y_{n_0}) < f(\overline{y}, \overline{y}) = \overline{y},$$

which is impossible. □

6.5.2 The Case $q < 1$

The main result in this section is the following:

Theorem 6.5.4 *Assume that*

$$q < 1.$$

Then the positive equilibrium of Eq(6.23) is globally asymptotically stable.

Proof. Clearly

$$y_{n+1} = \frac{p + qy_{n-1}}{1 + y_n} < p + qy_{n-1}, \quad n \geq 0. \tag{6.26}$$

Set

$$M = \frac{p}{1-q} + \epsilon,$$

where ϵ is some positive number.

From Eq(6.26) and Eq(6.23), it now follows that every solution of Eq(6.23) lies eventually in the interval $[0, M]$.

Set

$$f(x,y) = \frac{p+qy}{1+x}, \quad a = 0, \quad \text{and} \quad b = \frac{p}{1-q} + \epsilon.$$

Then clearly,

$$a \leq f(x,y) \leq b,$$

for all $(x,y) \in [a,b] \times [a,b]$. By applying Theorem 1.4.6 it follows that $\overline{y}$ is a global attractor of all solutions of Eq(6.23). The local stability of $\overline{y}$ was established in Theorem 6.5.1. The proof is complete. □

6.5.3 The Case $q > 1$

In this case we will show that there exist solutions that have two subsequences, one of which is monotonically approaching 0 and the other is monotonically approaching ∞. In fact this is true for every solution $\{y_n\}_{n=-1}^{\infty}$ whose initial conditions are such that the following inequalities are true:

$$y_{-1} < q-1 \quad \text{and} \quad y_0 > q-1+\frac{p}{q-1}.$$

To this end observe that

$$y_1 = \frac{p+qy_{-1}}{1+y_0} < \frac{p+q(q-1)}{1+y_0} < q-1$$

and

$$y_2 = \frac{p+qy_0}{1+y_1} > \frac{p+qy_0}{q} = y_0 + \frac{p}{q}$$

and by induction,

$$y_{2n-1} < q-1 \quad \text{and} \quad y_{2n} > y_0 + n\frac{p}{q} \quad \text{for} \quad n \geq 1.$$

Hence

$$\lim_{n\to\infty} y_{2n} = \infty$$

and

$$\lim_{n\to\infty} y_{2n+1} = \lim_{n\to\infty} \frac{p+qy_{2n-1}}{1+y_{2n}} = 0$$

and the proof is complete.

6.6 The Case $\beta = A = 0:\ x_{n+1} = \frac{\alpha+\gamma x_{n-1}}{Bx_n+Cx_{n-1}}$

This is the $(2,2)$-type Eq(6.6) which by the change of variables

$$x_n = \frac{\gamma}{C} y_n$$

reduces to the difference equation

$$y_{n+1} = \frac{p + y_{n-1}}{q y_n + y_{n-1}}, \quad n = 0, 1, \ldots \tag{6.27}$$

where

$$p = \frac{\alpha C}{\gamma^2} \quad \text{and} \quad q = \frac{B}{C}.$$

(Eq(6.27) was investigated in [45].)

Eq(6.27) has the unique positive equilibrium

$$\overline{y} = \frac{1 + \sqrt{1 + 4p(1+q)}}{2(1+q)}.$$

The linearized equation associated with Eq(6.27) about $\overline{y}$ is

$$z_{n+1} + \frac{q}{1+q} z_n - \frac{q\overline{y} - p}{(\overline{y} + p)(1+q)} z_{n-1} = 0, \quad n = 0, 1, \ldots . \tag{6.28}$$

By employing Theorem 1.1.1 we see that $\overline{y}$ is locally asymptotically stable when

$$(q-1)\overline{y} < 2p \tag{6.29}$$

and unstable (a saddle point) when

$$(q-1)\overline{y} > 2p. \tag{6.30}$$

Clearly the equilibrium $\overline{y}$ is the positive solution of the quadratic equation

$$(1+q)\overline{y}^2 - \overline{y} - p = 0.$$

If we now set

$$F(u) = (1+q)u^2 - u - p,$$

it is easy to see that (6.29) is satisfied if and only if either

$$q \leq 1$$

or

$$q > 1 \quad \text{and} \quad F\left(\frac{2p}{q-1}\right) > 0.$$

Similarly (6.30) is satisfied if and only if

$$q > 1 \quad \text{and} \quad F\left(\frac{2p}{q-1}\right) < 0.$$

The following result is now a consequence of the above discussion and some simple calculations:

Theorem 6.6.1 *The equilibrium $\overline{y}$ of Eq(6.27) is locally asymptotically stable when*

$$q < 1 + 4p \tag{6.31}$$

and unstable, and more precisely a saddle point equilibrium, when

$$q > 1 + 4p. \tag{6.32}$$

6.6.1 Existence and Local Stability of Period-Two Cycles

Let

$$\ldots, \phi, \psi, \phi, \psi, \ldots$$

be a period-two cycle of Eq(6.27). Then

$$\phi = \frac{p+\phi}{q\psi+\phi} \quad \text{and} \quad \psi = \frac{p+\psi}{q\phi+\psi}.$$

It now follows after some calculations that the following result is true:

Theorem 6.6.2 *Eq(6.27) has a prime period-two solution*

$$\ldots, \phi, \psi, \phi, \psi, \ldots$$

if and only if

$$q > 1 + 4p.$$

Furthermore when (6.32) holds, the period-two solution is "unique" and the values of ϕ and ψ are the positive roots of the quadratic equation

$$t^2 - t + \frac{p}{q-1} = 0.$$

To investigate the local stability of the two cycle

$$\ldots, \phi, \psi, \phi, \psi, \ldots$$

we set

$$u_n = y_{n-1} \quad \text{and} \quad v_n = y_n, \quad \text{for } n = 0, 1, \ldots$$

and write Eq(6.27) in the equivalent form:

$$u_{n+1} = v_n$$

$$v_{n+1} = \tfrac{p+u_n}{qv_n+u_n}, \quad n = 0, 1, \ldots .$$

Let T be the function on $(0, \infty) \times (0, \infty)$ defined by:

$$T\begin{pmatrix} u \\ v \end{pmatrix} = \begin{pmatrix} v \\ \frac{p+u}{qv+u} \end{pmatrix}.$$

Then

$$\begin{pmatrix} \phi \\ \psi \end{pmatrix}$$

is a fixed point of T^2, the second iterate of T. By a simple calculation we find that

$$T^2\begin{pmatrix} u \\ v \end{pmatrix} = \begin{pmatrix} g(u,v) \\ h(u,v) \end{pmatrix}$$

where

$$g(u,v) = \frac{p+u}{qv+u} \quad \text{and} \quad h(u,v) = \frac{p+v}{qg(u,v)+v}.$$

Clearly the two cycle is locally asymptotically stable when the eigenvalues of the Jacobian matrix J_{T^2}, evaluated at $\begin{pmatrix} \phi \\ \psi \end{pmatrix}$ lie inside the unit disk.

We have

$$J_{T^2}\begin{pmatrix} \phi \\ \psi \end{pmatrix} = \begin{pmatrix} \frac{\partial g}{\partial u}(\phi,\psi) & \frac{\partial g}{\partial v}(\phi,\psi) \\ \frac{\partial h}{\partial u}(\phi,\psi) & \frac{\partial h}{\partial v}(\phi,\psi) \end{pmatrix}$$

where,

$$\begin{aligned} \frac{\partial g}{\partial u}(\phi,\psi) &= \frac{q\psi - p}{(q\psi+\phi)^2}, \\ \frac{\partial g}{\partial v}(\phi,\psi) &= -\frac{(p+\phi)q}{(q\psi+\phi)^2}, \end{aligned}$$

$$\frac{\partial h}{\partial u}(\phi,\psi) = -\frac{q(p+\psi)(q\psi-p)}{(q\phi+\psi)^2(q\psi+\phi)^2},$$

$$\frac{\partial h}{\partial v}(\phi,\psi) = \frac{q^2(p+\phi)(p+\psi)}{(q\phi+\psi)^2(q\psi+\phi)^2} + \frac{q\phi-p}{(q\phi+\psi)^2}.$$

Set

$$\mathcal{S} = \frac{\partial g}{\partial u}(\phi,\psi) + \frac{\partial h}{\partial v}(\phi,\psi)$$

and

$$\mathcal{D} = \frac{\partial g}{\partial u}(\phi,\psi)\frac{\partial h}{\partial v}(\phi,\psi) - \frac{\partial g}{\partial v}(\phi,\psi)\frac{\partial h}{\partial u}(\phi,\psi).$$

Then it follows from Theorem 1.1.1 that both eigenvalues of $J_{T^2}\begin{pmatrix}\phi\\ \psi\end{pmatrix}$ lie inside the unit disk $|\lambda| < 1$, if and only if

$$|\mathcal{S}| < 1 + \mathcal{D} < 2. \tag{6.33}$$

Inequality (6.33) is equivalent to the following three inequalities:

$$\mathcal{S} < 1 + \mathcal{D} \tag{6.34}$$

$$-1 - \mathcal{D} < \mathcal{S} \tag{6.35}$$

$$\mathcal{D} < 1. \tag{6.36}$$

First we will establish Inequality (6.34).

To this end observe that (6.34) is equivalent to

$$(q\psi-p)(q\phi+\psi)^2 + (q\phi-p)(q\psi+\phi)^2 + q^2(p+\psi)(p+\phi) - (q\psi-p)(q\phi-p)$$
$$< (q\psi+\phi)^2(q\phi+\psi)^2$$

which is true if and only if

$$\frac{p}{q-1}(q^3+2q^2-4qp+2q^2p+2p-3q) + q - p + q^2p^2 + qp - p^2 < (q\psi+\phi)^2(q\phi+\psi)^2$$

which is true if and only if

$$-q^3p + 7q^2p - 8qp^2 + 4q^2p^2 + 4p^2 - 7qp + 2q^2 - q + p - q^3 < 0$$

which is true if and only if

$$-(p+1)(q-1)(q - \frac{p}{p+1})[q - (1+4p)] < 0$$

which is true because $q > 1 + 4p$.

Next we will establish Inequality (6.35). Observe that (6.35) is equivalent to

$$-(q\psi+\phi)^2(q\phi+\psi)^2$$
$$< (q\psi-p)(q\phi+\psi)^2+(q\phi-p)^2(q\psi+\phi)^2+q^2(p+\psi)(p+\phi)+(q\psi-p)(q\phi-p)$$

which is true if and only if

$$-(q\psi+\phi)^2(q\phi+\psi)^2 < \frac{p}{q-1}(q^3+4q^2-4qp+2q^2p+2p-3q)-p+q^2p^2$$

which is true if and only if

$$q^2+2pq+2p^2q^2-p^2+qp^2-q^3-3q^3p-2q^3p^2-p<0$$

which is true if and only if

$$p^2(-2q^3+2q^2+q-1)+p(-3q^3+2q-1)-q^3+q^2<0$$

which is true because $q>1$.

Finally, we establish Inequality (6.36). Observe that (6.36) is true if and only if

$$(q\psi-p)(q\phi-p)<(q\psi+\phi)^2(q\phi+\psi)^2$$

which is true if and only if

$$\frac{p}{q-1}[q^2-(q-1)(q-p)]<(pq-p+q)^2$$

which is true if and only if

$$q(pq-p+q)(pq-2p+q-1)>0$$

which is true because $q>1+4p$.

6.6.2 Invariant Intervals

Here we show that the interval I with end points one and $\frac{p}{q}$ is an invariant interval for Eq(6.27). More precisely we show that the values of any solution $\{y_n\}_{n=-1}^{\infty}$ of Eq(6.27) either remain forever outside the interval I or the solution is eventually trapped into I.

Theorem 6.6.3 *Let $\{y_n\}_{n=-1}^{\infty}$ be a solution of Eq(6.27). Then the following statements are true.*

(a) Assume that $p\le q$ and that for some $N\ge 0$,

$$\frac{p}{q}\le y_N\le 1.$$

Then

$$\frac{p}{q}\le y_n\le 1 \quad \textit{for all} \quad n\ge N.$$

(b) *Assume that $p \geq q$ and that for some $N \geq 0$,*

$$1 \leq y_N \leq \frac{p}{q}.$$

Then

$$1 \leq y_n \leq \frac{p}{q} \quad \textit{for all} \quad n \geq N.$$

Proof.

(a) Indeed we have

$$y_{N+1} = \frac{p + y_{N-1}}{qy_N + y_{N-1}} \leq \frac{p + y_{N-1}}{p + y_{N-1}} = 1$$

and

$$y_{N+1} = \frac{p + y_{N-1}}{qy_N + y_{N-1}} = \frac{p + y_{N-1}}{\frac{q}{p}(py_N + \frac{p}{q}y_{N-1})} \geq \left(\frac{p}{q}\right)\frac{p + y_{N-1}}{p + y_{N-1}} = \frac{p}{q}$$

and the proof follows by induction.

(b) Clearly,

$$y_{N+1} = \frac{p + y_{N-1}}{qy_N + y_{N-1}} \geq \frac{p + y_{N-1}}{p + y_{N-1}} = 1$$

and

$$y_{N+1} = \frac{p + y_{N-1}}{qy_N + y_{N-1}} = \left(\frac{p}{q}\right)\frac{p + y_{N-1}}{py_N + \frac{p}{q}y_{N-1}} \leq \left(\frac{p}{q}\right)\frac{p + y_{N-1}}{p + y_{N-1}} = \frac{p}{q}$$

and the proof follows by induction.

□

6.6.3 Semicycle Analysis

Here we present a thorough semicycle analysis of the solutions of Eq(6.27) relative to the equilibrium $\bar{y}$ and relative to the end points of the invariant interval of Eq(6.27).

Let $\{y_n\}_{n=-1}^{\infty}$ be a solution of Eq(6.27). Then observe that the following identities are true:

$$y_{n+1} - 1 = q\frac{\frac{p}{q} - y_n}{qy_n + y_{n-1}} \quad \text{for } n \geq 0, \tag{6.37}$$

$$y_{n+1} - \frac{p}{q} = \frac{qy_{n-1}(1 - \frac{p}{q}) + pq(1 - y_n)}{q(qy_n + y_{n-1})} \quad \text{for } n \geq 0, \tag{6.38}$$

$$y_n - y_{n+2} = \frac{qy_{n-1}(y_n - \frac{p}{q}) + (y_n - 1)(y_{n-1}y_n + qy_n^2)}{q(p + y_{n-1}) + y_n(qy_n + y_{n-1})} \quad \text{for } n \geq 0. \tag{6.39}$$

When

$$p = q$$

that is for the difference equation

$$y_{n+1} = \frac{p + y_{n-1}}{py_n + y_{n-1}}, \quad n = 0, 1, \ldots \tag{6.40}$$

the above identities reduce to the following:

$$y_{n+1} - 1 = p\frac{1 - y_n}{py_n + y_{n-1}} \quad \text{for } n \geq 0 \tag{6.41}$$

and

$$y_n - y_{n+2} = (y_n - 1)\frac{py_{n-1} + y_{n-1}y_n + py_n^2}{p(p + y_{n-1}) + y_n(py_n + y_{n-1})} \quad \text{for } n \geq 0. \tag{6.42}$$

The following three lemmas are now direct consequences of (6.37)-(6.42).

Lemma 6.6.1 *Assume that*

$$p > q$$

and let $\{y_n\}_{n=-1}^{\infty}$ be a solution of Eq(6.27). Then the following statements are true:

(i) If for some $N \geq 0$, $y_N < \frac{p}{q}$, then $y_{N+1} > 1$;

(ii) If for some $N \geq 0$, $y_N = \frac{p}{q}$, then $y_{N+1} = 1$;

(iii) If for some $N \geq 0$, $y_N > \frac{p}{q}$, then $y_{N+1} < 1$;

(iv) If for some $N \geq 0$, $y_N \geq 1$, then $y_{N+1} < \frac{p}{q}$;

(v) If for some $N \geq 0$, $y_N \geq \frac{p}{q}$, then $y_{N+2} < y_N$;

(vi) If for some $N \geq 0$, $y_N \leq 1$, then $y_{N+2} > y_N$;

(vii) If for some $N \geq 0$, $y_N \leq \frac{p}{q}$, then $y_n \in [1, \frac{p}{q}]$ for $n \geq N$. In particular $[1, \frac{p}{q}]$ is an invariant interval for Eq(6.27);

(viii)

$$1 < \overline{y} < \frac{p}{q}.$$

Lemma 6.6.2 *Assume that*

$$p = q$$

and let $\{y_n\}_{n=-1}^{\infty}$ be a solution of Eq(6.27). Then the following statements are true:

(i) If for some $N \geq 0$, $y_N < 1$, then $y_{N+1} > 1$;

(ii) *If for some* $N \geq 0$, $y_N = 1$, *then* $y_{N+1} = 1$;

(iii) *If for some* $N \geq 0$, $y_N > 1$, *then* $y_{N+1} < 1$;

(iv) *If for some* $N \geq 0$, $y_N < 1$, *then* $y_N < y_{N+2} < 1$;

(v) *If for some* $N \geq 0$, $y_N > 1$, *then* $y_N > y_{N+2} > 1$.

Lemma 6.6.3 *Assume that*

$$p < q$$

and let $\{y_n\}_{n=-1}^{\infty}$ *be a solution of Eq(6.27). Then the following statements are true:*

(i) *If for some* $N \geq 0$, $y_N < \frac{p}{q}$, *then* $y_{N+1} > 1$;

(ii) *If for some* $N \geq 0$, $y_N = \frac{p}{q}$, *then* $y_{N+1} = 1$;

(iii) *If for some* $N \geq 0$, $y_N > \frac{p}{q}$, *then* $y_{N+1} < 1$;

(iv) *If for some* $N \geq 0$, $y_N \leq 1$, *then* $y_{N+1} > \frac{p}{q}$;

(v) *If for some* $N \geq 0$, $y_N \leq \frac{p}{q}$, *then* $y_{N+1} \geq 1$;

(vi) *If for some* $N \geq 0$, $y_N \geq 1$, *then* $y_{N+2} < y_N$;

(vii) *If for some* $N \geq 0$, $y_N \leq \frac{p}{q}$, *then* $y_{N+2} > y_N$;

(viii) *If for some* $N \geq 0$, $y_N \leq 1$, *then* $y_n \in [\frac{p}{q}, 1]$ *for* $n \geq N$. *In particular* $[\frac{p}{q}, 1]$ *is an invariant interval for Eq(6.27);*

(ix)

$$\frac{p}{q} < \overline{y} < 1.$$

Note that the function

$$f(u, v) = \frac{p + v}{qu + v}$$

is always decreasing in u but in v it is decreasing when $u < \frac{p}{q}$ and increasing when $u > \frac{p}{q}$.

The following result is now a consequence of Theorems 1.7.1 and 1.7.2 and Lemmas 6.6.1-6.6.3:

Theorem 6.6.4 *Let $\{y_n\}_{n=-1}^{\infty}$ be a solution of Eq(6.27). Let I be the closed interval with end points 1 and $\frac{p}{q}$ and let J and K be the intervals which are disjoint from I and such that*

$$I \cup J \cup K = (0, \infty).$$

Then either all the even terms of the solution lie in J and all odd terms lie in K, or vice-versa, or for some $N \geq 0$,

$$y_n \in I \quad \textit{for } n \geq N. \tag{6.43}$$

When (6.43) holds, except for the length of the first semicycle of the solution, if

$$p < q$$

the length is one, while if

$$p > q$$

the length is at most two.

6.6.4 Global Behavior of Solutions

Recall that Eq(6.27) has a unique equilibrium $\overline{y}$ which is locally stable when (6.31) holds. When (6.32) holds, and only then, Eq(6.27) has a "unique" prime period-two solution

$$\ldots, \phi, \psi, \phi, \psi, \ldots \tag{6.44}$$

with

$$\phi + \psi = 1 \quad \text{and} \quad \phi\psi = \frac{p}{q-1}. \tag{6.45}$$

Also recall that a solution $\{y_n\}_{n=-1}^{\infty}$ of Eq(6.27) either eventually enters the interval I with end points 1 and $\frac{p}{q}$ and remains there, or stays forever outside I.

Now when

$$q \leq 1 + 4p$$

there are no period-two solutions and so the solutions must eventually enter and remain in I and, in view of Theorems 1.4.6 and 1.4.7, must converge to $\overline{y}$.

On the other hand when (6.32) holds, any solution which remains forever outside the interval $I = [\frac{p}{q}, 1]$ must converge to the two cycle (6.44)-(6.45). But this two cycle lies inside the interval I. Hence, when (6.32) holds, every solution of Eq(6.27) eventually enters and remains in the interval I.

The character of these solutions remains an **open question**.

The above observations are summarized in the following theorem:

Theorem 6.6.5 *(a) Assume*

$$q \leq 1 + 4p. \tag{6.46}$$

Then the equilibrium $\overline{y}$ of Eq(6.27) is global attractor.

(b) *Assume*

$$q > 1 + 4p.$$

Then every solution of Eq(6.27) eventually enters and remains in the interval $[\frac{p}{q}, 1]$.

6.7 The Case $\alpha = C = 0:\ x_{n+1} = \frac{\beta x_n + \gamma x_{n-1}}{A + Bx_n}$

This is the (2, 2)-type equation Eq(6.7) which by the change of variables

$$x_n = \frac{A}{B} y_n$$

reduces to the difference equation

$$y_{n+1} = \frac{py_n + qy_{n-1}}{1 + y_n}, \quad n = 0, 1, \ldots \tag{6.47}$$

where

$$p = \frac{\beta}{A} \quad \text{and} \quad q = \frac{\gamma}{A}.$$

(Eq(6.47) was investigated in [52].)
The equilibrium points of Eq(6.47) are the solutions of the equation

$$\overline{y} = \frac{p\overline{y} + q\overline{y}}{1 + \overline{y}}.$$

So $\overline{y} = 0$ is always an equilibrium point, and when

$$p + q > 1,$$

$\overline{y} = p + q - 1$ is the only positive equilibrium point of Eq(6.47).

The following result follows from Theorem 1.3.1 and the linearized stability Theorem 1.1.1:

Theorem 6.7.1 (a) *Assume*

$$p + q \leq 1.$$

Then the zero equilibrium of Eq(6.47) is globally asymptotically stable.

(b) *Assume*

$$p + q > 1.$$

Then the zero equilibrium of Eq(6.47) is unstable. More precisely it is a saddle point when

$$1 - p < q < 1 + p$$

and a repeller when

$$q > 1 + p.$$

(c) *Assume*

$$1 - p < q < 1 + p.$$

Then the positive equilibrium $\overline{y} = p + q - 1$ *of Eq(6.47) is locally asymptotically stable.*

(d) *Assume*

$$q > 1 + p.$$

Then the positive equilibrium $\overline{y} = p + q - 1$ *of Eq(6.47) is an unstable saddle point.*

Concerning prime period-two solutions it follows from Sections 2.5 that the following result is true:

Theorem 6.7.2 *Eq(6.47) has a prime period-two solution*

$$\ldots, \phi, \psi, \phi, \psi, \ldots$$

if and only if

$$q = 1 + p. \tag{6.48}$$

Furthermore when (6.48) holds, the values of ϕ *and* ψ *of all prime period-two solutions are given by:*

$$\{\phi, \psi \in (p, \infty) : \psi = \frac{p\phi}{\phi - p} \neq 2p\}.$$

6.7.1 Invariant Intervals

The main result here is the following theorem about Eq(6.47) with $q \neq 1$. The case

$$q = 1 \tag{6.49}$$

is treated in Section 6.7.3.

Theorem 6.7.3 (a) *Assume that*

$$q < 1. \tag{6.50}$$

Then every solution of Eq(6.47) eventually enters the interval $(0, \frac{p}{q}]$. *Furthermore,* $(0, \frac{p}{q}]$ *is an invariant interval for Eq(6.47). That is, every solution of Eq(6.47) with initial conditions in* $(0, \frac{p}{q}]$, *remains in this interval.*

(b) *Assume that*

$$q > 1. \tag{6.51}$$

Then every solution of Eq(6.47) eventually enters the interval $[\frac{p}{q}, \infty)$. *Furthermore,* $[\frac{p}{q}, \infty)$ *is an invariant interval for Eq(6.47).*

The proof of the above theorem is an elementary consequence of the following lemma, whose proof is straightforward and will be omitted.

Lemma 6.7.1 *(a) Assume that (6.50) holds. Then for any $k, m \in \mathbf{N}$,*

$$y_k \leq \frac{p}{q} \Longrightarrow y_{k+2} \leq p < \frac{p}{q}$$

and

$$y_{k+2} \geq \frac{p}{q^m} \Longrightarrow y_k > \frac{p}{q^{m+1}} > p.$$

(b) Assume that (6.51) holds. Then for any $k, m \in \mathbf{N}$,

$$y_k \geq \frac{p}{q} \Longrightarrow y_{k+2} \geq p > \frac{p}{q}$$

and

$$y_{k+2} \leq \frac{p}{q^m} \Longrightarrow y_k > \frac{p}{q^{m+1}} > p.$$

6.7.2 Semicycle Analysis of Solutions When $q \leq 1$

Here we discuss the behavior of the semicycles of solutions of Eq(6.47) when

$$p + q > 1 \quad \text{and} \quad q \leq 1.$$

Hereafter, $\overline{y}$ refers to the positive equilibrium $p + q - 1$ of Eq(6.47).

Let $f : [0, \infty) \times [0, \infty) \to [0, \infty)$ denote the function

$$f(x, y) = \frac{px + qy}{1 + x}.$$

Then f satisfies the negative feedback condition; that is,

$$(x - \overline{x})(f(x, x) - x) < 0 \quad \text{for} \quad x \in (0, \infty) - \{\overline{x}\}.$$

Consider the function $g : [0, \infty) \to [0, \infty)$ defined by

$$g(x) = \frac{(p + q)x}{1 + x}.$$

Observe that $g(\overline{y}) = \overline{y}$ and that g increases on $[0, \infty)$.

Theorem 6.7.4 *Suppose*

$$p + q > 1 \quad \textit{and} \quad q \leq 1.$$

Let $\{y_n\}_{n=-1}^{\infty}$ be a positive solution of Eq(6.47). Then the following statements are true for every $n \geq 0$:

(a) If $\overline{y} \le y_{n-1}, y_n,$ *then* $\overline{y} \le y_{n+1} \le \max\{y_{n-1}, y_n\}.$

(b) If $y_{n-1}, y_n \le \overline{y},$ *then* $\min\{y_{n-1}, y_n\} \le y_{n+1} < \overline{y}.$

(c) If $y_{n-1} < \overline{y} \le y_n,$ *then* $y_{n-1} < y_{n+1} < y_n.$

(d) If $y_n < \overline{y} \le y_{n-1},$ *then* $y_n < y_{n+1} < y_{n-1}.$

(e) If $\overline{y} < y_k < \frac{\overline{y}}{2-q},$ *for every* $k \ge -1$*, then* $\{y_n\}$ *decreases to the equilibrium* $\overline{y}$*.*

Proof.

(a) Suppose $\overline{y} \le y_{n-1}, y_n$. Then

$$y_{n+1} = \frac{py_n + qy_{n-1}}{1+y_n} \le \frac{(p+q)\max\{y_{n-1}, y_n\}}{1+\overline{y}} = \max\{y_{n-1}, y_n\}.$$

Since the function

$$\frac{px + q\overline{x}}{1+x},$$

is increasing for $\overline{y} < \frac{p}{q}$, which is equivalent to $q \le 1$, we obtain

$$y_{n+1} = \frac{py_n + qy_{n-1}}{1+y_n} \ge \frac{py_n + q\overline{y}}{1+y_n} \ge g(\overline{y}) = \overline{y}.$$

Observe that all inequalities are strict if we assume that $y_{n-1} > \overline{y}$ or $y_n > \overline{y}$.

(b) Suppose $y_{n-1}, y_n \le \overline{y}$. Then in view of the previous case

$$y_{n+1} = \frac{py_n + qy_{n-1}}{1+y_n} \le \frac{py_n + q\overline{y}}{1+y_n} < g(\overline{y}) = \overline{y}$$

and

$$y_{n+1} = \frac{py_n + qy_{n-1}}{1+y_n} \ge \frac{(p+q)\min\{y_{n-1}, y_n\}}{1+\overline{y}} = \min\{y_{n-1}, y_n\}.$$

Observe that all inequalities are strict if we assume that $y_{n-1} < \overline{y}$ or $y_n < \overline{y}$.

(c) Suppose $y_{n-1} < \overline{y} \le y_n$. Then

$$y_{n+1} = \frac{py_n + qy_{n-1}}{1+y_n} < \frac{(p+q)y_n}{1+y_n} = \frac{(1+\overline{y})y_n}{1+y_n} < y_n.$$

Since $q \le 1$ and $y_{n-1} < \overline{y}$,

$$p - qy_{n-1} > p - q\overline{y} = (1-q)(p+q) \ge 0.$$

Thus $\frac{px+qy_{n-1}}{1+x}$ increases in x and so

$$y_{n+1} = \frac{py_n + qy_{n-1}}{1+y_n} > \frac{py_{n-1} + qy_{n-1}}{1+y_{n-1}} = g(y_{n-1}) > y_{n-1}.$$

The last inequality follows from the negative feedback property.

(*d*) Suppose $y_n < \overline{y} \leq y_{n-1}$. Then

$$y_{n+1} = \frac{py_n + qy_{n-1}}{1+y_n} > \frac{py_n + qy_n}{1+y_n} = g(y_n) > y_n$$

by the negative feedback property.

To see that $y_{n+1} < y_{n-1}$, we must consider two cases:
Case 1. Suppose $p - qy_{n-1} \geq 0$. Then $\frac{px+qy_{n-1}}{1+x}$ increases in x and so

$$y_{n+1} = \frac{py_n + qy_{n-1}}{1+y_n} \leq \frac{py_{n-1} + qy_{n-1}}{1+y_{n-1}} < y_{n-1}.$$

The last inequality follows from the negative feedback property.

Case 2. Suppose $p - qy_{n-1} < 0$. Then

$$y_{n+1} = \frac{py_n + qy_{n-1}}{1+y_n} < \frac{qy_{n-1}(y_n+1)}{1+y_n} \leq y_{n-1}.$$

(*e*) Observe that

$$y_{n+1} = \frac{py_n + qy_{n-1}}{1+y_n} > \overline{y} = p + q - 1$$

and so

$$y_{n-1} \geq qy_{n-1} > \overline{y} + (q-1)y_n > y_n$$

from which the result follows.

□

6.7.3 Semicycle Analysis and Global Attractivity When $q = 1$

Here we discuss the behavior of the solutions of Eq(6.47) when

$$q = 1.$$

Here the positive equilibrium of Eq(6.47) is

$$\overline{y} = p.$$

Theorem 6.7.5 *Suppose that $q = 1$ and let $y_{-1} + y_0 > 0$. Then the equilibrium $\overline{y}$ of Eq(6.47) is globally asymptotically stable. More precisely the following statements are true.*

(a) Assume $y_{-1} \geq p$ and $y_0 \geq p$. Then $y_n \geq p$ for all $n > 0$ and

$$\lim_{n\to\infty} y_n = p. \tag{6.52}$$

(b) Assume $y_{-1} \leq p$ and $y_0 \leq p$. Then $y_n \leq p$ for all $n > 0$ and (6.52) holds.

(c) Assume either $y_{-1} < p < y_0$ or $y_{-1} > p > y_0$. Then $\{y_n\}_{n=-1}^{\infty}$ oscillates about the equilibrium p with semicycles of length one and (6.52) holds.

Proof. *(a)* The case where $y_{-1} = y_0 = p$ is trivial. We will assume that $y_{-1} \geq p$ and $y_0 > p$. The case $y_{-1} > p$ and $y_0 \geq p$ is similar and will be omitted. Then

$$y_1 = \frac{py_0 + y_{-1}}{1 + y_0} \geq \frac{py_0 + p}{1 + y_0} = p,$$

and

$$y_2 = \frac{py_1 + y_0}{1 + y_1} > \frac{py_1 + p}{1 + y_1} = p.$$

It follows by induction that $y_n \geq p$ for all $n > 0$. Furthermore, $y_{-1} \geq p$ implies that

$$py_0 + y_{-1} \leq y_{-1}y_0 + y_{-1}$$

and so

$$y_1 = \frac{py_0 + y_{-1}}{1 + y_0} \leq \frac{y_{-1}y_0 + y_{-1}}{1 + y_0} = y_{-1}.$$

Similarly, we can show that $y_2 < y_0$ and by induction

$$y_{-1} \geq y_1 \geq y_3 \geq \ldots \geq p$$

and

$$y_0 > y_2 > y_4 > \ldots > p.$$

Thus, there exist $m \geq p$ and $M \geq p$ such that

$$\lim_{k\to\infty} y_{2k+1} = m \quad \text{and} \quad \lim_{k\to\infty} y_{2k} = M.$$

As Eq(6.47) has no period-two solutions when $q = 1$, it follows that $m = M = p$ and the proof of part *(a)* is complete.

It is interesting to note that

$$y_{-1} = p \Rightarrow y_{2k+1} = p \text{ for } k \geq 0 \quad \text{and} \quad y_0 = p \Rightarrow y_{2k} = p \text{ for } k \geq 0.$$

(b) The proof of *(b)* is similar to the proof of *(a)* and will be omitted.

(c) Assume $y_{-1} < p < y_0$. The proof when $y_0 < p < y_{-1}$ is similar and will be omitted.

$$y_1 = \frac{py_0 + y_{-1}}{1 + y_0} < \frac{py_0 + p}{1 + y_0} = p,$$

and

$$y_2 = \frac{py_1 + y_0}{1 + y_1} > \frac{py_1 + p}{1 + y_1} = p.$$

It follows by induction that the solution oscillates about p with semicycles of length one. Next, we claim that the subsequence $\{y_{2k+1}\}_{k=-1}^{\infty}$ is increasing, and so convergent. In fact, since $y_{-1} + py_0 > y_{-1} + y_{-1}y_0$ then

$$y_1 = \frac{py_0 + y_{-1}}{1 + y_0} > y_{-1}$$

and the claim follows by induction. In a similar way we can show that the subsequence $\{y_{2k}\}_{k=0}^{\infty}$ is decreasing, and so convergent. Thus, there exist $m \leq p$ and $M \geq p$ such that

$$\lim_{k\to\infty} y_{2k+1} = m \quad \text{and} \quad \lim_{k\to\infty} y_{2k} = M$$

and as in *(a)*, $m = M$. □

6.7.4 Global Attractivity When $q < 1$

Here we consider the case where

$$1 - p < q < 1,$$

and we show that the equilibrium point $\overline{y} = p + q - 1$ of Eq(6.47) is global attractor of all positive solutions of Eq(6.47).

Theorem 6.7.6 *Assume*

$$1 - p < q < 1.$$

Then every positive solution of Eq(6.47) converges to the positive equilibrium $\overline{y}$.

Proof. Observe that the consistency condition $\overline{y} \leq \frac{p}{q}$ is equivalent to the condition $q \leq 1$. By Theorem 6.7.3 *(a)* we may assume that the initial conditions y_{-1} and y_0 lie in the interval $I = (0, \frac{p}{q}]$. Now clearly the function

$$f(x, y) = \frac{px + qy}{1 + x}$$

satisfies the hypotheses of Theorem 1.4.5 in the interval I from which the result follows. □

6.7.5 Long-term Behavior of Solutions When $q > 1$

Here we discuss the behavior of the solutions of Eq(6.47) when

$$q > 1$$

which is equivalent to

$$\overline{y} > \frac{p}{q}.$$

In this case, every positive solution of Eq(6.47) lies eventually in the interval $[p, \infty)$. Concerning the long-term behavior of solutions of Eq(6.47) we may assume, without loss of generality, that

$$y_{-1}, y_0 \in [p, \infty).$$

Then the change of variable

$$y_n = u_n + p,$$

transforms Eq(6.47) to the equation

$$u_{n+1} = \frac{p(q-1) + qu_{n-1}}{1 + p + u_n} \tag{6.53}$$

where $u_n \geq 0$ for $n \geq 0$.

The character of solutions of Eq(6.53) was investigated in Section 6.5. By applying the results of Section 6.5 to Eq(6.53), we obtain the following theorems about the solutions of Eq(6.47) when $q > 1$:

Theorem 6.7.7 *After the first semicycle, an oscillatory solution of Eq(6.47) oscillates about the positive equilibrium $\overline{y}$ with semicycles of length one.*

Theorem 6.7.8 *(a) Assume*

$$q = 1 + p.$$

Then every solution of Eq(6.47) converges to a period-two solution.

(b) Assume

$$1 < q < 1 + p.$$

Then every positive solution of Eq(6.47) converges to the positive equilibrium of Eq(6.47).

(c) Assume

$$q > 1 + p.$$

Then Eq(6.47) has unbounded solutions.

6.8 The Case $\alpha = B = 0: \; x_{n+1} = \frac{\beta x_n + \gamma x_{n-1}}{A + C x_{n-1}}$

This is the $(2,2)$-type Eq(6.8) which by the change of variables

$$x_n = \frac{\gamma}{C} y_n$$

reduces to the difference equation

$$y_{n+1} = \frac{p y_n + y_{n-1}}{q + y_{n-1}}, \quad n = 0, 1, \ldots \tag{6.54}$$

where

$$p = \frac{\beta}{\gamma} \quad \text{and} \quad q = \frac{A}{\gamma}.$$

(Eq(6.54) was investigated in [53].)

The equilibrium points of Eq(6.54) are the solutions of the equation

$$\overline{y} = \frac{p\overline{y} + \overline{y}}{q + \overline{y}}.$$

So $\overline{y} = 0$ is always an equilibrium point of Eq(6.54) and when

$$p + 1 > q,$$

Eq(6.54) also possesses the unique positive equilibrium $\overline{y} = p + 1 - q$.

The following result follows from Theorems 1.1.1 and 1.3.1.

Theorem 6.8.1 *(a) Assume*

$$p + 1 \le q.$$

Then the zero equilibrium of Eq(6.54) is globally asymptotically stable.

(b) Assume

$$p + 1 > q. \tag{6.55}$$

Then the zero equilibrium of Eq(6.54) is unstable and the positive equilibrium $\overline{y} = p + 1 - q$ of Eq(6.54) is locally asymptotically stable. Furthermore the zero equilibrium is a saddle point when

$$1 - p < q < 1 + p$$

and a repeller when

$$q < 1 - p.$$

In the remainder of this section we will investigate the character of the positive equilibrium of Eq(6.54) and so we will assume without further mention that (6.55) holds. Our goal is to show that when (6.55) holds and $y_{-1} + y_0 > 0$, the positive equilibrium $\overline{y} = p + 1 - q$ of Eq(6.54) is globally asymptotically stable.

6.8.1 Invariant Intervals and Semicycle Analysis

Let $\{y_n\}_{n=-1}^{\infty}$ be a positive solution of Eq(6.54). Then the following identities are easily established:

$$y_{n+1} - 1 = p\frac{y_n - \frac{q}{p}}{q + y_{n-1}} \quad \text{for } n \geq 0, \tag{6.56}$$

$$y_{n+1} - \frac{q}{p} = \frac{p^2[y_n - \left(\frac{q}{p}\right)^2] + (p-q)y_{n-1}}{p(q + y_{n-1})} \quad \text{for } n \geq 0, \tag{6.57}$$

and

$$y_{n+1} - y_n = \frac{(p-q)y_n + (1 - y_n)y_{n-1}}{q + y_{n-1}} \quad \text{for } n \geq 0. \tag{6.58}$$

Note that the positive equilibrium

$$\overline{y} = p + 1 - q$$

of Eq(6.54) is such that:

$$\overline{y} \text{ is } \begin{cases} < 1 & if \;\; p < q < p+1 \\ = 1 & if \;\; p = q \\ > 1 & if \;\; p > q. \end{cases}$$

When

$$p = q$$

that is for the difference equation

$$y_{n+1} = \frac{py_n + y_{n-1}}{p + y_{n-1}}, \quad n = 0, 1, \ldots \tag{6.59}$$

the identities (6.56)-(6.58) reduce to the following:

$$y_{n+1} - 1 = (y_n - 1)\frac{p}{p + y_{n-1}} \quad \text{for } n \geq 0 \tag{6.60}$$

and

$$y_{n+1} - y_n = (1 - y_n)\frac{y_{n-1}}{p + y_{n-1}} \quad \text{for } n \geq 0. \tag{6.61}$$

The following three lemmas are now direct consequences of the above identities.

Lemma 6.8.1 *Assume that*

$$p < q \tag{6.62}$$

and let $\{y_n\}_{n=-1}^{\infty}$ *be a positive solution of Eq(6.54). Then the following statements are true:*

(i) *If for some* $N \geq 0$, $y_N < \frac{q}{p}$, *then* $y_{N+1} < 1$;

(ii) *If for some* $N \geq 0$, $y_N = \frac{q}{p}$, *then* $y_{N+1} = 1$;

(iii) *If for some* $N \geq 0$, $y_N > \frac{q}{p}$, *then* $y_{N+1} > 1$;

(iv) *If for some* $N \geq 0$, $y_N > \left(\frac{q}{p}\right)^2$, *then* $y_{N+1} < \frac{q}{p}$;

(v) *If for some* $N \geq 0$, $y_N \leq 1$, *then* $y_{N+1} < 1$;

(vi) *If for some* $N \geq 0$, $y_N \geq 1$, *then* $y_{N+1} < y_N$.

Lemma 6.8.2 *Assume that*

$$p = q \tag{6.63}$$

and let $\{y_n\}_{n=-1}^{\infty}$ *be a positive solution of Eq(6.54). Then the following statements are true:*

(i) *If* $y_0 < 1$, *then for* $n \geq 0$, $y_n < 1$ *and the solution is strictly increasing;*

(ii) *If* $y_0 = 1$, *then* $y_n = 1$ *for* $n \geq 0$;

(iii) *If* $y_0 > 1$, *then for* $n \geq 0$, $y_n > 1$ *and the solution is strictly decreasing.*

Lemma 6.8.3 *Assume that*

$$p > q \tag{6.64}$$

and let $\{y_n\}_{n=-1}^{\infty}$ *be a positive solution of Eq(6.54). Then the following statements are true:*

(i) *If for some* $N \geq 0$, $y_N < \frac{q}{p}$, *then* $y_{N+1} < 1$;

(ii) *If for some* $N \geq 0$, $y_N = \frac{q}{p}$, *then* $y_{N+1} = 1$;

(iii) *If for some* $N \geq 0$, $y_N > \frac{q}{p}$, *then* $y_{N+1} > 1$;

(iv) *If for some* $N \geq 0$, $y_N > \left(\frac{q}{p}\right)^2$, *then* $y_{N+1} > \frac{q}{p}$;

(v) *If for some* $N \geq 0$, $y_N \leq 1$, *then* $y_{N+1} > y_N$;

(vi) *If for some* $N \geq 0$, $y_N \geq 1$, *then* $y_{N+1} > \frac{q}{p}$ *and* $y_{N+2} > 1$.

6.8.2 Global Stability of the Positive Equilibrium

When (6.62) holds, it follows from Lemma 6.8.1 that if a solution $\{y_n\}_{n=-1}^{\infty}$ is such that

$$y_n \geq 1 \quad \text{for all} \quad n \geq 0$$

then the solution decreases and its limit lies in the interval $[1, \infty)$. This is impossible because $\overline{y} < 1$. Hence, by Lemma 6.8.1, every positive solution of Eq(6.54) eventually enters and remains in the interval $(0, 1)$. Now in the interval $(0, 1)$ the function

$$f(u, v) = \frac{pu + v}{q + v}$$

is increasing in both arguments and by Theorem 1.4.8, $\overline{y}$ is a global attractor.

When (6.63) holds, it follows from Lemma 6.8.2 that every solution of Eq(6.54) converges to the equilibrium $\overline{y} = 1$.

Next, assume that (6.64) holds. Here it is clear from Lemma 6.8.3 that every solution of Eq(6.54) eventually enters and remains in the interval $(1, \infty)$. Without loss of generality we will assume that

$$y_n > 1 \quad \text{for} \quad n \geq -1.$$

Set

$$y_n = 1 + (q + 1)u_n \quad \text{for} \quad n \geq -1.$$

Then we can see that $\{u_n\}_{n=-1}^{\infty}$ satisfies the difference equation

$$u_{n+1} = \frac{\frac{p-q}{(q+1)^2} + \frac{p}{q+1}u_n}{1 + u_{n-1}}, \quad n = 0, 1, \ldots \tag{6.65}$$

with positive parameters and positive initial conditions. This (2,2)-type equation was investigated in Section 6.3 where we established in Theorem 6.3.3 that every solution converges to its positive equilibrium

$$\overline{u} = \frac{p - q}{q + 1}.$$

From the above observations and in view of Theorem 6.8.1 we have the following result:

Theorem 6.8.2 *Assume that $p + 1 > q$ and that $y_{-1} + y_0 > 0$. Then the positive equilibrium $\overline{y} = p + 1 - q$ of Eq(6.54) is globally asymptotically stable.*

6.9 The Case $\alpha = A = 0:\ x_{n+1} = \frac{\beta x_n + \gamma x_{n-1}}{Bx_n + Cx_{n-1}}$

This is the (2, 2)-type Eq(6.9) which by the change of variables

$$x_n = \frac{\gamma}{C} y_n$$

reduces to the equation

$$y_{n+1} = \frac{py_n + y_{n-1}}{qy_n + y_{n-1}}, \quad n = 0, 1, \ldots \tag{6.66}$$

where

$$p = \frac{\beta}{\gamma} \quad \text{and} \quad q = \frac{B}{C}.$$

To avoid a degenerate situation we will also assume that

$$p \neq q.$$

(Eq(6.66) was investigated in [54].)

The only equilibrium point of Eq(6.66) is

$$\overline{y} = \frac{p+1}{q+1}$$

and the linearized equation of Eq(6.66) about $\overline{y}$ is

$$z_{n+1} - \frac{p-q}{(p+1)(q+1)} z_n + \frac{p-q}{(p+1)(q+1)} z_{n-1} = 0, \quad n = 0, 1, \ldots .$$

By applying the linearized stability Theorem 1.1.1 we obtain the following result:

Theorem 6.9.1 *(a) Assume that*

$$p > q.$$

Then the positive equilibrium of Eq(6.66) is locally asymptotically stable.

(b) Assume that

$$p < q.$$

Then the positive equilibrium of Eq(6.66) is locally asymptotically stable when

$$q < pq + 1 + 3p \tag{6.67}$$

and is unstable (and more precisely a saddle point equilibrium) when

$$q > pq + 1 + 3p. \tag{6.68}$$

6.9.1 Existence of a Two Cycle

It follows from Section 2.5 that when $p > q$, Eq(6.66) has no prime period-two solutions. On the other hand when

$$p < q$$

and

$$q > pq + 1 + 3p,$$

Eq(6.66) possesses a unique prime period-two solution:

$$\ldots, \phi, \psi, \phi, \psi, \ldots \tag{6.69}$$

where the values of ϕ and ψ are the (positive and distinct) solutions of the quadratic equation

$$t^2 - (1-p)t + \frac{p(1-p)}{q-1} = 0. \tag{6.70}$$

In order to investigate the stability nature of this prime period-two solution, we set

$$u_n = y_{n-1} \quad \text{and} \quad v_n = y_n \quad \text{for } n = 0, 1, \ldots$$

and write Eq(6.66) in the equivalent form:

$$\begin{aligned} u_{n+1} &= v_n \\ v_{n+1} &= \tfrac{pv_n+u_n}{qv_n+u_n}, \quad n = 0, 1, \ldots. \end{aligned}$$

Let T be the function on $(0, \infty) \times (0, \infty)$ defined by:

$$T\begin{pmatrix} u \\ v \end{pmatrix} = \begin{pmatrix} v \\ \frac{pv+u}{qv+u} \end{pmatrix}.$$

Then

$$\begin{pmatrix} \phi \\ \psi \end{pmatrix}$$

is a fixed point of T^2, the second iterate of T. One can see that

$$T^2\begin{pmatrix} u \\ v \end{pmatrix} = \begin{pmatrix} g(u,v) \\ h(u,v) \end{pmatrix}$$

where

$$g(u,v) = \frac{pv+u}{qv+u} \quad \text{and} \quad h(u,v) = \frac{p\frac{pv+u}{qv+u} + v}{q\frac{pv+u}{qv+u} + v}.$$

The prime period-two solution (6.69) is asymptotically stable if the eigenvalues of the Jacobian matrix J_{T^2}, evaluated at $\begin{pmatrix} \phi \\ \psi \end{pmatrix}$ lie inside the unit disk. One can see that

$$J_{T^2}\begin{pmatrix} \phi \\ \psi \end{pmatrix} = \begin{pmatrix} -\frac{(p-q)\psi}{(q\psi+\phi)^2} & \frac{(p-q)\phi}{(q\psi+\phi)^2} \\ \\ -\frac{(p-q)^2\psi^2}{(q\psi+\phi)^2(q\phi+\psi)^2} & \frac{(p-q)\phi}{(q\phi+\psi)^2}\left(\frac{(p-q)\psi}{(q\psi+\phi)^2} - 1\right) \end{pmatrix}.$$

Set

$$P = \frac{p-q}{(q\psi+\phi)^2} \quad \text{and} \quad Q = \frac{p-q}{(q\phi+\psi)^2}.$$

Then

$$J_{T^2}\begin{pmatrix}\phi\\ \psi\end{pmatrix} = \begin{pmatrix}-P\psi & P\phi\\ -PQ\psi^2 & PQ\phi\psi - Q\phi\end{pmatrix}$$

and its characteristic equation is

$$\lambda^2 + (P\psi + Q\phi - PQ\phi\psi)\lambda + PQ\phi\psi = 0.$$

By applying Theorem 1.1.1 *(c)*, it follows that both eigenvalues of

$$J_{T^2}\begin{pmatrix}\phi\\ \psi\end{pmatrix}$$

lie inside the unit disk if and only if

$$|P\psi + Q\phi - PQ\phi\psi| < 1 + PQ\phi\psi < 2$$

or equivalently if and only if the following three inequalities hold:

$$P\psi + Q\phi + 1 > 0 \tag{6.71}$$

$$P\psi + Q\phi < 1 + 2PQ\phi\psi \tag{6.72}$$

and

$$PQ\phi\psi < 1. \tag{6.73}$$

Observe that

$$\phi > 0, \psi > 0, P < 0 \quad \text{and} \quad Q < 0$$

and so Inequality (6.72) is always true.

Next we will establish Inequality (6.71). We will use the fact that

$$\phi + \psi = 1 - p, \quad \phi\psi = \frac{p(1-p)}{q-1}$$

$$\phi = \frac{p\psi + \phi}{q\psi + \phi} \quad \text{and} \quad \psi = \frac{p\phi + \psi}{q\phi + \psi}.$$

Inequality (6.71) is equivalent to

$$\frac{(p-q)\psi}{(q\psi+\phi)^2} + \frac{(p-q)\phi}{(q\phi+\psi)^2} + 1 > 0$$

which is true if and only if

$$(q-p)[(q\phi+\psi)^2\psi + (q\psi+\phi)^2\phi] < (q\psi+\phi)^2(q\phi+\psi)^2$$

if and only if

$$(q-p)[(q\phi+\psi)(p\phi+\psi)+(q\psi+\phi)(p\psi+\phi)] < [(q\phi+\psi)(q\psi+\phi)]^2. \tag{6.74}$$

Now observe that the lefthand side of (6.74) is

$$\begin{aligned} I &= (q-p)[(qp+1)(\phi^2+\psi^2)+2(p+q)\phi\psi] \\ &= (q-p)[(qp+1)(\phi+\psi)^2 - 2\phi\psi(qp+1-p-q)] \\ &= (q-p)[(qp+1)(1-p)^2 - 2\frac{p(1-p)}{q-1}(p-1)(q-1)] \\ &= (q-p)(1-p)^2(qp+1+2p). \end{aligned}$$

The righthand side of (6.74) is

$$\begin{aligned} II &= [(q\psi+\phi)(q\phi+\psi)]^2 \\ &= [(q^2+1)\psi\phi + q(\phi^2+\psi^2)]^2 \\ &= [q(\phi+\psi)^2 + (q-1)^2\phi\psi]^2 \\ &= [q(1-p)^2 + (q-1)p(1-p)]^2 \\ &= (1-p)^2(q-p)^2. \end{aligned}$$

Hence, Inequality (6.71) is true if and only if (6.68) holds.

Finally, Inequality (6.73) is equivalent to

$$(p-q)^2\phi\psi < (q\psi+\phi)^2(q\phi+\psi)^2$$

which is true if and only if

$$(q-p)\sqrt{\phi\psi} < (q\psi+\phi)(q\phi+\psi)$$

if and only if

$$(q-p)\sqrt{\phi\psi} < (q^2+1)\phi\psi + q(\phi^2+\psi^2)$$

if and only if

$$(q-p)\sqrt{\phi\psi} < (q^2+1)\phi\psi + q[(\phi+\psi)^2 - 2\phi\psi]$$

if and only if

$$(q-p)\sqrt{\phi\psi} < (q-1)^2\phi\psi + q(\phi+\psi)^2$$

if and only if

$$(q-p)\sqrt{\phi\psi} < (q-1)^2\frac{p(1-p)}{q-1} + q(1-p)^2$$

if and only if

$$(q-p)\sqrt{\phi\psi} < (1-p)(q-p)$$

if and only if

$$\frac{p(1-p)}{q-1} < (1-p)^2$$

if and only if $q > pq + 1$ which is clearly true.

In summary, the following result is true about the local stability of the prime period-two solution (6.69) of Eq(6.66):

Theorem 6.9.2 *Assume that Condition (6.68) holds. Then Eq(6.66) posseses the prime period-two solution*

$$\ldots, \phi, \psi, \phi, \psi, \ldots$$

where ϕ and ψ are the two positive and distinct roots of the quadratic Eq(6.70). Furthermore this prime period-two solution is locally asymptotically stable.

6.9.2 Semicycle Analysis

In this section, we present a semicycle analysis of the solutions of Eq(6.66).

Theorem 6.9.3 *Let $\{y_n\}$ be a nontrivial solution of Eq(6.66). Then the following statements are true:*

(a) Assume $p > q$. Then $\{y_n\}$ oscillates about the equilibrium $\overline{y}$ with semicycles of length two or three, except possibly for the first semicycle which may have length one. The extreme in each semicycle occurs in the first term if the semicycle has two terms and in the second term if the semicycle has three terms.

(b) Assume $p < q$. Then either $\{y_n\}$ oscillates about the equilibrium $\overline{y}$ with semicycles of length one, after the first semicycle, or $\{y_n\}$ converges monotonically to $\overline{y}$.

Proof. *(a)* The proof follows from Theorem 1.7.4 by observing that the condition $p > q$ implies that the function

$$f(x, y) = \frac{px + y}{qx + y}$$

is increasing in x and decreasing in y. This function also satisfies Condition (1.35).

(b) The proof follows from Theorem 1.7.1 by observing that when $p < q$, the function $f(x, y)$ is increasing in y and decreasing in x. □

6.9.3 Global Stability Analysis When $p < q$

The main result in this section is the following

Theorem 6.9.4 *Assume that*

$$p < q$$

and (6.67) holds. Then the positive equilibrium $\overline{y}$ of Eq(6.66) is globally asymptotically stable.

Proof. Set

$$f(x,y)=\frac{px+y}{qx+y},$$

and note that $f(x,y)$ is decreasing in x for each fixed y, and increasing in y for each fixed x. Also clearly,

$$\frac{p}{q}\leq f(x,y)\leq 1 \quad \text{for all } x,y>0.$$

Finally in view of (6.67), Eq(6.66) has no prime period-two solution. Now the conclusion of Theorem 6.9.4 follows as a consequence of Theorem 1.4.6 and the fact that $\overline{y}$ is locally asymptotically stable. □

The method employed in the proof of Theorem 1.4.6 can also be used to establish that certain solutions of Eq(6.66) converge to the two cycle (6.69) when instead of (6.67), (6.68) holds.

Theorem 6.9.5 *Assume that (6.68) holds. Let $\phi,\psi,\phi,\psi,\ldots$, with $\phi<\psi$, denote the two cycle of Eq(6.66). Assume that for some solution $\{y_n\}_{n=-1}^{\infty}$ of Eq(6.66) and for some index $N\geq -1$,*

$$y_N\geq\psi \quad \textit{and} \quad y_{N+1}\leq\phi. \tag{6.75}$$

Then this solution converges to the two cycle $\phi,\psi,\phi,\psi,\ldots$.

Proof. Assume that (6.75) holds. Set

$$f(x,y)=\frac{px+y}{qx+y}.$$

Then clearly,

$$y_{N+2}=f(y_{N+1},y_N)\geq f(\phi,\psi)=\psi,$$

$$y_{N+3}=f(y_{N+2},y_{N+1})\leq f(\psi,\phi)=\phi$$

and in general

$$y_{N+2k}\geq\psi \quad \text{and} \quad y_{N+2k+1}\leq\phi \quad \text{for } k=0,1,\ldots.$$

Now as in the proof of Theorem 1.4.6

$$\limsup_{n\to\infty} y_n=\psi \quad \text{and} \quad \liminf_{n\to\infty} y_n=\phi$$

from which we conclude that

$$\lim_{k\to\infty} y_{N+2k}=\psi \quad \text{and} \quad \lim_{k\to\infty} y_{N+2k+1}=\phi.$$

□

Next we want to find more cases where the conclusion of the last theorem holds. We consider first the case where eventually consecutive terms y_i, y_{i+1} lie between m and M.

Lemma 6.9.1 *Suppose that for some $i \geq 0$,*

$$m \leq y_i, y_{i+1} \leq M.$$

Then $y_k \in [m, M]$ for every $k > i$.

Proof. Using the monotonic character of the function f we have

$$m = f(M, m) \leq f(y_{i+1}, y_i) = y_{i+2} \leq f(m, M) = M,$$

and the result follows by induction. □

If we never get two successive terms in the interval $[m, M]$ and if we never get two successive terms outside of the interval $[m, M]$ (that is, for every i, either $y_i > M$ and $y_{i+1} < m$ or $y_i < m$ and $y_{i+1} > M$), then we have one of the following four cases:

(A) $y_{2n} > M$ and $M \geq y_{2n+1} > m$ for every n;
(B) $y_{2n+1} > M$ and $M \geq y_{2n} > m$ for every n;
(C) $y_{2n} < m$ and $m \leq y_{2n+1} < M$ for every n;
(D) $y_{2n+1} < m$ and $m \leq y_{2n} < M$ for every n.

Lemma 6.9.2 *In all cases (A)-(D) the corresponding solution $\{y_n\}$ converges to the two cycle*

$$\ldots \phi, \psi, \phi, \psi, \ldots .$$

Proof. Since the proofs are similar for all cases we will give the details in case *(A)*. By our earlier observation we have that $M \geq \limsup_{i\to\infty} y_i$, and so

$$M \geq \limsup_{n\to\infty} y_{2n} \geq \liminf_{n\to\infty} y_{2n} \geq M.$$

Hence $\lim_{n\to\infty} y_{2n} = M$. From Eq(6.66) we have

$$y_{2n+1} = \frac{y_{2n} - y_{2n}y_{2n+2}}{qy_{2n+2} - p}.$$

and so

$$\lim_{n\to\infty} y_{2n+1} = \frac{M - M^2}{qM - p}.$$

On the other hand by solving the equation

$$M = f(m, M) = \frac{pm + M}{qm + M}$$

for m we find

$$m = \frac{M - M^2}{qM - p}.$$

Thus, $\lim_{n\to\infty} y_{2n+1} = m$ and the proof is complete. □

Since the case in which the solution lies outside of the interval $[m, M]$ is answered by Theorem 6.9.5, we are left with the case in which the solution lies eventually in $[m, M]$.

Lemma 6.9.3 *Suppose $M \geq y_{i+2} \geq y_i \geq \overline{y} \geq y_{i+1} \geq y_{i+3} \geq m$ and not both y_{i+2} and y_{i+3} are equal to $\overline{y}$. Then $\{y_n\}$ converges to the two cycle*

$$\ldots \phi, \psi, \phi, \psi, \ldots.$$

Proof. Here we have

$$y_{i+4} = f(y_{i+3}, y_{i+2}) \geq f(y_{i+1}, y_i) = y_{i+2}$$

and

$$y_{i+5} = f(y_{i+4}, y_{i+3}) \leq f(y_{i+2}, y_{i+1}) = y_{i+3}.$$

Induction on i then establishes the monotonicity of the two sequences. Thus, the sequence $\{y_{i+2k}\}$ is non decreasing and bounded above by M, and the sequence $\{y_{i+2k+1}\}$ is a non increasing sequence and bounded below by m. Therefore, both subsequences converge and their limits form a two cycle. In view of the uniqueness of the two cycle and the fact that at least one of y_i, y_{i+1} is different from $\overline{y}$, the proof is complete.The case $M \geq y_{i+3} \geq y_{i+1} \geq \overline{y} \geq y_i \geq y_{i+2} \geq m$ is handled in a similar way. □

Using a similar technique one can prove that if

$$M \geq y_i \geq y_{i+2} \geq \overline{y} \geq y_{i+3} \geq y_{i+1} \geq m$$

or

$$M \geq y_{i+1} \geq y_{i+3} \geq \overline{y} \geq y_{i+2} \geq y_i \geq m$$

we would get convergence to $\overline{y}$, but this situation cannot occur, as the following results show. Let us consider the first case. The second case can be handled in a similar way. In this case we need to know more about the condition $y_{i+2} = f(y_{i+1}, y_i) \leq y_i$, so we are led to the general equation

$$f(x, y) = y. \tag{6.76}$$

It makes sense to consider this equation only for $y \in (\frac{p}{q}, 1)$, in which case Eq(6.76) has a unique positive solution for x, which of course depends on y. We denote this value by $y^\star$. Thus $f(y^\star, y) = y$. The correspondence between y and $y^\star$ has the following properties:

Lemma 6.9.4 *(1) $f(x, y) > y$ if and only if $x < y^\star$.*

(2) $x > y$ if and only if $x^\star < y^\star$.

(3) For $\overline{y} < y < M$, $f(y, y^\star) < y^\star$ and $y > y^{\star\star}$, and for $m < y < \overline{y}$, $f(y, y^\star) > y^\star$ and $y^{\star\star} > y$.

Proof.

(1) Since f is decreasing in the first variable and $f(y^\star, y) = y$, $x < y^\star$ if and only if $f(x, y) > f(y^\star, y) = y$.

(2) Solving $f(y^\star, y) = y$ for $y^\star$ gives

$$y^\star = g(y) = \frac{y - y^2}{qy - p}.$$

Computing the derivative of g we get

$$g'(y) = -\frac{p - 2y + qy^2}{(p - qy)^2}.$$

The discriminant of the numerator is $4p^2 - 4pq = 4p(p - q) < 0$, so the roots of the numerator are not real. For $y \in (\frac{p}{q}, 1)$, $g'(y) < 0$, so $y^\star$ is decreasing function of y.

(3) Eliminating $y^\star$ from the equations $y = f(y^\star, y)$ and $y^\star = f(y, y^\star)$ we get

$$\frac{x - x^2}{qx - p} = \frac{px + \frac{x - x^2}{qx - p}}{qx + \frac{x - x^2}{qx - p}}.$$

Simplifying this equation we get a fourth degree polynomial equation with roots $0, \overline{y}, m, M$. Thus each of these roots is simple and so $f(y, y^\star) - y^\star$ changes sign as y passes each of these values. Clearly as y approaches $\frac{p}{q}$ $y^\star$ approaches ∞. However $\frac{p}{q} < f(y, y^\star) < 1$. Thus for $y < m$, $f(y, y^\star) < y^\star$. Hence

$$m < y < \overline{y} \Rightarrow f(y, y^\star) > y^\star,$$

$$\overline{y} < y < M \Rightarrow f(y, y^\star) < y^\star,$$

$$M < y <\Rightarrow f(y, y^\star) > y^\star.$$

By *(1)* then, for $m < y < \overline{y}$, $\quad y < y^{\star\star}$, and for $\overline{y} < y < M$, $\quad y > y^{\star\star}$.

□

Now we will use this result to prove the following:

Lemma 6.9.5 *Assume that for some i,*

$$M \geq y_i \geq y_{i+2} \geq \overline{y} \geq y_{i+1} \geq m.$$

Then $y_{i+3} \leq y_{i+1}$ with equality only in the case of a two cycle.

Proof. To show $y_{i+3} = f(y_{i+2}, y_{i+1}) \leq y_{i+1}$, we must show $y_{i+2} \geq y^{\star}_{i+1}$.

Since $y_{i+2} = f(y_{i+1}, y_i) \leq y_i$, we have $y_{i+1} \geq y^{\star}_i$. Thus by *(2)* of the previous lemma $y^{\star\star}_i \geq y^{\star}_{i+1}$, and by *(3)* $y_i \geq y^{\star\star}_i$. Thus $y_i \geq y^{\star}_{i+1}$. Since f is increasing in the second argument and $y_i \geq y^{\star}_{i+1}$ we get

$$y_{i+2} = f(y_{i+1}, y_i) \geq f(y_{i+1}, y^{\star}_{i+1}).$$

By *(3)* of the previous lemma, since $m \leq y_{i+1} \leq \overline{y}$, $f(y_{i+1}, y^{\star}_{i+1}) \geq y^{\star}_{i+1}$. Thus $y_{i+2} \geq y^{\star}_{i+1}$, as required. To get equality we must have $y_{i+1} = y^{\star}_i$ or equivalently $y_{i+2} = y_i$, as well as $y_{i+3} = y_{i+1}$. Thus equality occurs only in the case of a two cycle. □

The case

$$m \leq y_i \leq y_{i+2} \leq \overline{y} \leq y_{i+1} \leq M$$

is similar.

We also obtain the following lemma:

Lemma 6.9.6 *Unless $\{y_n\}$ is a period-two solution, there is no i such that either*

$$M \geq y_i \geq y_{i+2} \geq \ldots \geq \overline{y} \geq y_{i+1} \geq y_{i+3} \geq \ldots \geq m,$$

or

$$M \geq \ldots \geq y_{i+2} \geq y_i \geq \overline{y} \geq \ldots \geq y_{i+3} \geq y_{i+1} \geq \ldots \geq m.$$

Proof. We will consider the first case; the second case is similar. Since $\{y_{i+2k+1}\}$ is a bounded monotonic sequence it has a limit c. The value c will be one value in a two cycle, so either $c = \overline{y}$ or $c = m$. If $c = \overline{y}$, then $y_{i+2n+1} = \overline{y}$ for all n, and solving for y_{i+2n} we see $y_{i+2n} = \overline{y}$ also. But then $y_i = y_{i+1} = \overline{y}$, and our solution is a two cycle.. If $c = m$, solving for $\lim_{n\to\infty} y_{i+2n}$ gives $\lim_{n\to\infty} y_{i+2n} = M$. This can only happen if $y_{i+2n} = M$ for all n, in which case $y_{i+2n+1} = m$ for all n, which also gives a two cycle. □

The above sequence of lemmas leads to the following result:

Theorem 6.9.6 *Assume that*

$$p < q,$$

and that Eq(6.66) possesses the two-cycle solution (6.69). Then every oscillatory solution of Eq(6.66) converges to this two cycle.

Remark. Since some of the oscillatory solutions of Eq(6.66) are generated by initial values $\{y_{-1}, y_0\}$ that are on opposite sides of the equilibrium $\overline{y}$, it follows by Theorem 6.9.6 that all such pairs belong to the stable manifold of the two cycle.

6.9.4 Global Stability Analysis When $p > q$

Here we have the following result:

Theorem 6.9.7 *Assume that*

$$p > q$$

and

$$p \leq pq + 1 + 3q. \tag{6.77}$$

Then the positive equilibrium $\overline{y}$ of Eq(6.66) is globally asymptotically stable.

Proof. We will employ Theorem 1.4.5. To this end, set

$$f(x, y) = \frac{px + y}{qx + y},$$

and observe that when $p > q$, the function $f(x, y)$ is increasing in x for each fixed y, and is decreasing in y for each fixed x. Also

$$1 \leq f(x, y) \leq \frac{p}{q} \quad \text{for all } x, y > 0.$$

Finally observe that when (6.77) holds, the only solution of the system

$$M = \frac{pM + m}{qM + m}, \qquad m = \frac{pm + M}{qm + M},$$

is

$$m = M.$$

Now the result is a consequence of Theorem 1.4.5. □

6.10 Open Problems and Conjectures

Open Problem 6.10.1 (See [64]) *Assume that*

$$a, b, c, d \in R.$$

(a) Investigate the forbidden set $\mathcal{F}$ of the difference equation

$$x_{n+1} = \frac{ax_n + bx_{n-1}}{cx_n + dx_{n-1}} x_n, \qquad n = 0, 1, \ldots .$$

(b) For every point $(x_{-1}, x_0) \notin \mathcal{F}$, investigate the asymptotic behavior and the periodic nature of the solution $\{x_n\}_{n=-1}^{\infty}$.

(See Section 1.6.)

Open Problem 6.10.2 *Assume that*

$$\alpha_n, \beta_n, A_n, B_n \in R \quad for \quad n = 0, 1, \ldots$$

are convergent sequences of real numbers with finite limits. Investigate the forbidden set and the asymptotic character of solutions of the nonautonomous Riccati equation

$$x_{n+1} = \frac{\alpha_n + \beta_n x_n}{A_n + B_n x_n}, \quad n = 0, 1, \ldots .$$

(See Section 1.6 for the autonomous case.)

Conjecture 6.10.1 *Assume*

$$p, q \in (0, \infty).$$

Show that every positive solution of Eq(6.10) has a finite limit.

(Note that by Theorem 6.3.3, we need only to confirm this conjecture for $p > 2(q+1)$ and $q \geq 1$. See also [36].)

Conjecture 6.10.2 *Assume that*

$$p, q \in (0, \infty).$$

Show that every positive solution of Eq(6.11) has a finite limit.

(Note that by Theorem 6.4.4, this conjecture has been confirmed for $q \leq 1 + 4p$.)

Conjecture 6.10.3 *Assume* $p \in (0, \infty)$. *Show that the equation*

$$y_{n+1} = \frac{p + y_{n-1}}{1 + y_n}, \quad n = 0, 1, \ldots$$

has a positive and monotonically decreasing solution.

(Note that by Theorem 6.5.2, every positive solution converges to a, not necessarily prime, period-two solution.)

Conjecture 6.10.4 *Assume that*

$$p, q \in (0, \infty).$$

Show that every positive solution of Eq(6.27) either converges to a finite limit or to a two cycle.

(See Section 6.6.)

Open Problem 6.10.3 *Assume that*

$$p, q \in (0, \infty)$$

and

$$q > 1 + 4p.$$

Find the basin of attraction of the period-two solution of Eq(6.27).

Open Problem 6.10.4 *Assume that*

$$p, q \in (0, \infty)$$

and

$$q > pq + 1 + 3p.$$

Find the basin of attraction of the period-two solution of Eq(6.66).

Conjecture 6.10.5 *Assume*

$$p > q.$$

Show that every positive solution of Eq(6.66) has a finite limit.

(Note that by Theorem 6.9.7, this conjecture has been confirmed for $p \leq pq + 1 + 3q$.)

Open Problem 6.10.5 *Find the set of all initial conditions $(y_{-1}, y_0) \in R \times R$ through which the solution of the equation*

$$y_{n+1} = \frac{y_n + y_{n-1}}{1 + y_n}$$

is well defined and converges to 1, as $n \to \infty$.

Open Problem 6.10.6 *Find the set of all initial conditions* $(y_{-1}, y_0) \in R \times R$ *through which the solution of the equation*

$$y_{n+1} = \frac{y_n + y_{n-1}}{1 + y_{n-1}}$$

is well defined and converges to 1, *as* $n \to \infty$.

Open Problem 6.10.7 *For each of the following difference equations determine the "good" set* $\mathcal{G} \subset R \times R$ *of initial conditions* $(y_{-1}, y_0) \in R \times R$ *through which the equation is well defined for all* $n \geq 0$. *Then for every* $(y_{-1}, y_0) \in \mathcal{G}$, *investigate the long-term behavior of the solution* $\{y_n\}_{n=-1}^{\infty}$:

$$y_{n+1} = \frac{1 - y_n}{1 - y_{n-1}} \tag{6.78}$$

$$y_{n+1} = \frac{y_n - 1}{y_n - y_{n-1}} \tag{6.79}$$

$$y_{n+1} = \frac{1 - y_{n-1}}{1 - y_n} \tag{6.80}$$

$$y_{n+1} = \frac{1 - y_{n-1}}{y_n - y_{n-1}} \tag{6.81}$$

$$y_{n+1} = \frac{y_n - y_{n-1}}{y_n - 1} \tag{6.82}$$

$$y_{n+1} = \frac{y_n - y_{n-1}}{1 - y_{n-1}} \tag{6.83}$$

$$y_{n+1} = \frac{y_n + y_{n-1}}{y_n - y_{n-1}}. \tag{6.84}$$

For those equations from the above list for which you were successful, extend your result by introducing arbitrary real parameters in the equation.

Open Problem 6.10.8 *Assume*

$$\alpha, \beta, \gamma, A, B, C \in (0, \infty).$$

Investigate the asymptotic behavior and the periodic nature of solutions of the difference equation

$$x_{n+1} = \frac{\alpha x_n + \beta x_{n-1} + \gamma x_{n-2}}{A x_n + B x_{n-1} + C x_{n-2}}, \quad n = 0, 1, \dots .$$

Extend and generalize.

Open Problem 6.10.9 *Assume that* $\{p_n\}_{n=0}^{\infty}$ *and* $\{q_n\}_{n=0}^{\infty}$ *are convergent sequences of nonnegative real numbers with finite limits,*

$$p = \lim_{n\to\infty} p_n \quad \text{and} \quad q = \lim_{n\to\infty} q_n.$$

Investigate the asymptotic behavior and the periodic nature of all positive solutions of each of the following eight difference equations:

$$y_{n+1} = \frac{p_n + y_n}{q_n + y_n}, \quad n = 0, 1, \ldots \tag{6.85}$$

$$y_{n+1} = \frac{p_n + q_n y_n}{1 + y_{n-1}}, \quad n = 0, 1, \ldots \tag{6.86}$$

$$y_{n+1} = \frac{y_n + p_n}{y_n + q_n y_{n-1}}, \quad n = 0, 1, \ldots \tag{6.87}$$

$$y_{n+1} = \frac{p_n + q_n y_{n-1}}{1 + y_n}, \quad n = 0, 1, \ldots \tag{6.88}$$

$$y_{n+1} = \frac{p_n + y_{n-1}}{q_n y_n + y_{n-1}}, \quad n = 0, 1, \ldots \tag{6.89}$$

$$y_{n+1} = \frac{p_n y_n + q_n y_{n-1}}{1 + y_n}, \quad n = 0, 1, \ldots \tag{6.90}$$

$$y_{n+1} = \frac{p_n y_n + y_{n-1}}{q_n + y_{n-1}}, \quad n = 0, 1, \ldots \tag{6.91}$$

$$y_{n+1} = \frac{p_n y_n + y_{n-1}}{q_n y_n + y_{n-1}}, \quad n = 0, 1, \ldots . \tag{6.92}$$

Open Problem 6.10.10 *Assume that* $\{p_n\}_{n=0}^{\infty}$ *and* $\{q_n\}_{n=0}^{\infty}$ *are period-two sequences of nonnegative real numbers. Investigate the global character of all positive solutions of Eqs(6.85)-(6.92). Extend and generalize.*

Open Problem 6.10.11 *Assume* $q \in (1, \infty)$.

(a) *Find the set* B *of all initial conditions* $(y_{-1}, y_0) \in (0, \infty) \times (0, \infty)$ *such that the solutions* $\{y_n\}_{n=-1}^{\infty}$ *of Eq(6.23) are bounded.*

(b) *Let* $(y_{-1}, y_0) \in B$. *Investigate the asymptotic behavior of* $\{y_n\}_{n=-1}^{\infty}$.

Open Problem 6.10.12 *Assume* $q \in (1, \infty)$.

(a) *Find the set* B *of all initial conditions* $(y_{-1}, y_0) \in (0, \infty) \times (0, \infty)$ *such that the solutions* $\{y_n\}_{n=-1}^{\infty}$ *of Eq(6.47) are bounded.*

(b) *Let* $(y_{-1}, y_0) \in B$. *Investigate the asymptotic behavior of* $\{y_n\}_{n=-1}^{\infty}$.

Open Problem 6.10.13 (A Plant-Herbivore System; See [9], [3], and [2]) *Assume*

$$\alpha \in (1, \infty), \quad \beta \in (0, \infty), \quad \text{and} \quad \gamma \in (0, 1) \quad \text{with} \quad \alpha + \beta > 1 + \frac{\beta}{\gamma}.$$

Obtain conditions for the global asymptotic stability of the positive equilibrium of the system

$$\left.\begin{aligned} x_{n+1} &= \tfrac{\alpha x_n}{\beta x_n + e^{y_n}} \\ y_{n+1} &= \gamma(x_n + 1) y_n \end{aligned}\right\}, n = 0, 1, \ldots . \tag{6.93}$$

Conjecture 6.10.6 *Assume* $x_0, y_0 \in (0, \infty)$ *and that*

$$\alpha \in (2, 3), \quad \beta = 1 \quad \text{and} \quad \gamma = \frac{1}{2}.$$

Show that the positive equilibrium of System (6.93) is globally asymptotically stable.

Open Problem 6.10.14 (Discrete Epidemic Models; See [14].) *Let* $A \in (0, \infty)$.

(a) *Determine the set* $\mathcal{G}$ *of initial conditions* $(y_{-1}, y_0) \in (0, \infty) \times (0, \infty)$ *through which the solutions* $\{x_n\}_{n=-1}^{\infty}$ *of the difference equation*

$$x_{n+1} = (1 - x_n - x_{n-1})(1 - e^{-Ax_n}), \quad n = 0, 1, \ldots \tag{6.94}$$

are nonnegative.

(b) *Let* $(x_{-1}, x_0) \in \mathcal{G}$. *Investigate the boundedness character, the periodic nature, and the asymptotic behavior of the solution* $\{x_n\}_{n=-1}^{\infty}$ *of Eq(6.94).*

(c) *Extend and generalize.*

Open Problem 6.10.15 (**The Flour Beetle Model**; See [48]). *Assume*

$$a \in (0,1), \quad b \in (0,\infty), \quad \textit{and} \quad c_1, c_2 \in [0,\infty) \quad \textit{with} \quad c_1 + c_2 > 0.$$

Obtain necessary and sufficient conditions for the global asymptotic stability of the Flour Beetle Model:

$$x_{n+1} = ax_n + bx_{n-2}e^{-c_1 x_n - c_2 x_{n-2}}, \quad n = 0, 1, \ldots$$

with positive initial conditions.

Open Problem 6.10.16 (**A Population Model**) *Assume*

$$\alpha \in (0,1) \quad \textit{and} \quad \beta \in (1,\infty).$$

Investigate the global character of all positive solutions of the system

$$\left.\begin{aligned} x_{n+1} &= \alpha x_n e^{-y_n} + \beta \\ y_{n+1} &= \alpha x_n (1 - e^{-y_n}) \end{aligned}\right\}, n = 0, 1, \ldots .$$

which may be viewed as a population model.

Open Problem 6.10.17 *Assume that*

$$p, q \in [0,\infty) \quad \textit{and} \quad k \in \{2, 3, \ldots\}.$$

Investigate the global behavior of all positive solutions of each of the following difference equations:

$$y_{n+1} = \frac{p + qy_n}{1 + y_{n-k}}, \quad n = 0, 1, \ldots \tag{6.95}$$

$$y_{n+1} = \frac{y_n + p}{y_n + qy_{n-k}}, \quad n = 0, 1, \ldots \tag{6.96}$$

$$y_{n+1} = \frac{p + qy_{n-k}}{1 + y_n}, \quad n = 0, 1, \ldots \tag{6.97}$$

$$y_{n+1} = \frac{p + y_{n-k}}{qy_n + y_{n-k}}, \quad n = 0, 1, \ldots \tag{6.98}$$

$$y_{n+1} = \frac{y_n + py_{n-k}}{y_n + q}, \quad n = 0, 1, \ldots \tag{6.99}$$

$$y_{n+1} = \frac{py_n + y_{n-k}}{q + y_{n-k}}, \quad n = 0, 1, \ldots \tag{6.100}$$

$$y_{n+1} = \frac{py_n + y_{n-k}}{qy_n + y_{n-k}}, \quad n = 0, 1, \ldots . \tag{6.101}$$

Open Problem 6.10.18 *Obtain a general result for an equation of the form*

$$y_{n+1} = f(y_n, y_{n-1}), \quad n = 0, 1, \ldots$$

which extends and unifies the global asymptotic stability results for Eqs(6.11) and (6.27).

Open Problem 6.10.19 *Assume that every positive solution of an equation of the form*

$$x_{n+1} = f(x_n, x_{n-1}), \quad n = 0, 1, \ldots \tag{6.102}$$

converges to a period-two solution. Obtain necessary and sufficient conditions on f so that every positive solution of the equation

$$x_{n+1} = f(x_{n-2}, x_{n-1}), \quad n = 0, 1, \ldots \tag{6.103}$$

also converges to a period-two solution.
Extend and generalize.

Open Problem 6.10.20 *Assume that Eq(6.102) has a two cycle which is locally asymptotically stable. Obtain necessary and sufficient conditions on f so that the same two cycle is a locally asymptotically stable solution of Eq(6.103).*

Chapter 7

(1, 3)-Type Equations

7.1 Introduction

Eq(1) contains the following three equations of the (1, 3)-type:

$$x_{n+1} = \frac{\alpha}{A + Bx_n + Cx_{n-1}}, \quad n = 0, 1, \ldots \tag{7.1}$$

$$x_{n+1} = \frac{\beta x_n}{A + Bx_n + Cx_{n-1}}, \quad n = 0, 1, \ldots \tag{7.2}$$

and

$$x_{n+1} = \frac{\gamma x_{n-1}}{A + Bx_n + Cx_{n-1}}, \quad n = 0, 1, \ldots . \tag{7.3}$$

Please recall our classification convention in which all the parameters in the above (1, 3)-type equations are positive and that the initial conditions are nonnegative.

7.2 The Case $\beta = \gamma = 0$: $x_{n+1} = \frac{\alpha}{A+Bx_n+Cx_{n-1}}$

This is the (1, 3)-type Eq.(7.1) which by the change of variables

$$x_n = \frac{\alpha}{A} y_n$$

reduces to the difference equation

$$y_{n+1} = \frac{1}{1 + py_n + qy_{n-1}}, \quad n = 0, 1, \ldots \tag{7.4}$$

where

$$p = \frac{\alpha B}{A^2} \quad \text{and} \quad q = \frac{\alpha C}{A^2}.$$

One can easily see that the positive equilibrium of Eq(7.4) is locally asymptotically stable for all values of the parameters and that Eq(7.4) has no prime period two solutions.

The following result is now a straightforward consequence of either the stability trichotomy Theorem 1.4.4, or Theorem 1.4.7 in the interval $[0, 1]$.

Theorem 7.2.1 *The positive equilibrium of Eq(7.4) is globally asymptotically stable.*

7.3 The Case $\alpha = \gamma = 0: \ x_{n+1} = \frac{\beta x_n}{A+Bx_n+Cx_{n-1}}$

This is the $(1, 3)$-type Eq(7.2) which by the change of variables

$$x_n = \frac{A}{C}y_n$$

reduces to the difference equation

$$y_{n+1} = \frac{y_n}{1 + py_n + qy_{n-1}}, \quad n = 0, 1, \ldots \tag{7.5}$$

where

$$p = \frac{\beta}{A} \quad \text{and} \quad q = \frac{B}{C}.$$

Eq(7.5) always has zero as an equilibrium point and when

$$p > 1$$

it also has the unique positive equilibrium point

$$\bar{y} = \frac{p-1}{q+1}.$$

The following result is a straightforward consequence of Theorem 1.3.1.

Theorem 7.3.1 *Assume*

$$p \leq 1.$$

Then the zero equilibrium of Eq(7.5) is globally asymptotically stable.

The following result is a straightforward consequence of Theorem 1.4.2. It also follows by applying Theorem 1.4.5 in the interval $[0, \frac{p}{q}]$.

Theorem 7.3.2 *Assume that $y_{-1}, y_0 \in (0, \infty)$ and*

$$p > 1.$$

Then the positive equilibrium of Eq(7.5) is globally asymptotically stable.

7.4 The Case $\alpha = \beta = 0:\ x_{n+1} = \frac{\gamma x_{n-1}}{A+Bx_n+Cx_{n-1}}$

This is the $(1,3)$-type Eq(7.3) which by the change of variables

$$x_n = \frac{\gamma}{C} y_n$$

reduces to the equation

$$y_{n+1} = \frac{y_{n-1}}{p + qy_n + y_{n-1}}, \quad n = 0, 1, \ldots \tag{7.6}$$

where

$$p = \frac{A}{\gamma} \quad \text{and} \quad q = \frac{B}{C}.$$

Eq(7.6) always has the zero equilibrium and when

$$p < 1$$

it also has the unique positive equilibrium

$$\overline{y} = \frac{1-p}{1+q}.$$

The subsequent result is a straightforward application of Theorems 1.1.1 and 1.3.1.

Theorem 7.4.1 *(a) The zero equilibrium of Eq(7.6) is globally asymptotically stable when*

$$p \geq 1$$

and is unstable (a repeller) when

$$p < 1.$$

(b) Assume that $p < 1$. Then the positive equilibrium of Eq(7.6) is locally asymptotically stable when

$$q < 1$$

and is unstable (a saddle point) when

$$q > 1.$$

Concerning prime period-two solutions one can see that Eq(7.6) has a prime period-two solution

$$\ldots, \phi, \psi, \phi, \psi, \ldots$$

if and only if

$$p < 1.$$

Furthermore when

$$p < 1 \quad \text{and} \quad q \neq 1$$

the only prime period-two solution is

$$\ldots, 0, 1-p, 0, 1-p, \ldots .$$

On the other hand, when

$$p < 1 \quad \text{and} \quad q = 1$$

all prime period-two solutions of Eq(7.6) are given by

$$\ldots, \phi, 1-p-\phi, \phi, 1-p-\phi, \ldots$$

with

$$0 \leq \phi \leq 1-p \quad \text{and} \quad \phi \neq \frac{1-p}{2}.$$

Theorem 7.4.2 *Assume that*

$$p < 1 \quad \textit{and} \quad q > 1. \tag{7.7}$$

Then the period-two solution

$$\ldots, 0, 1-p, 0, 1-p, \ldots \tag{7.8}$$

of Eq(7.6) is locally asymptotically stable.

Proof. It follows from Section 2.6 that here

$$J_{T^2}\begin{pmatrix} 0 \\ 1-p \end{pmatrix} = \begin{pmatrix} \frac{1}{p+q(1-p)} & 0 \\ -\frac{q(1-p)}{p+q(1-p)} & p \end{pmatrix}$$

and in view of (7.7) both eigenvalues of $J_{T^2}\begin{pmatrix} 0 \\ 1-p \end{pmatrix}$ lie in the disk $|\lambda| < 1$. Therefore the prime period-two solution (7.8) is locally asymptotically stable. □

7.5 Open Problems and Conjectures

Conjecture 7.5.1 *Assume*

$$p, q \in (0, 1).$$

Show that every positive solution of Eq(7.6) has a finite limit.

Conjecture 7.5.2 *Assume*

$$p < 1 \quad \text{and} \quad q \geq 1.$$

Show that every positive solution of Eq(7.6) converges to a, not necessarily prime, period-two solution.

Open Problem 7.5.1 *Assume*

$$p \in (0, 1) \quad \text{and} \quad q \in (1, \infty).$$

Find the basin of attraction of the two cycle

$$\ldots, 0, 1-p, 0, 1-p, \ldots$$

of Eq(7.6).

Open Problem 7.5.2 *Investigate the asymptotic character and the periodic nature of all positive solutions of the difference equation*

$$x_{n+1} = \frac{x_{n-2}}{1 + px_n + qx_{n-1} + x_{n-2}}, \qquad n = 0, 1, \ldots$$

where

$$p, q \in [0, \infty).$$

Extend and generalize.

Open Problem 7.5.3 *For each of the equations given below, find the set $\mathcal{G}$ of all points $(x_{-1}, x_0) \in R \times R$ through which the equation is well defined for all $n \geq 0$:*

$$x_{n+1} = \frac{1}{1 + x_n + x_{n-1}} \tag{7.9}$$

$$x_{n+1} = \frac{x_n}{1 + x_n + x_{n-1}} \tag{7.10}$$

$$x_{n+1} = \frac{x_{n-1}}{1 + x_n + x_{n-1}}. \tag{7.11}$$

Now for each of the equations (7.9)-(7.11), let $(x_{-1}, x_0) \in \mathcal{G}$ be an initial point through which the corresponding equation is well defined for all $n \geq 0$, and investigate the long term-behavior of the solution $\{x_n\}_{n=-1}^{\infty}$.

Open Problem 7.5.4 *Assume that* $\{p_n\}_{n=0}^{\infty}$ *and* $\{q_n\}_{n=0}^{\infty}$ *are convergent sequences of nonnegative real numbers with finite limits,*

$$p = \lim_{n\to\infty} p_n \quad \textit{and} \quad q = \lim_{n\to\infty} q_n.$$

Investigate the asymptotic behavior and the periodic nature of all positive solutions of each of the following three difference equations:

$$y_{n+1} = \frac{1}{1 + p_n y_n + q_n y_{n-1}}, \quad n = 0, 1, \ldots \tag{7.12}$$

$$y_{n+1} = \frac{y_n}{1 + p_n y_n + q_n y_{n-1}}, \quad n = 0, 1, \ldots \tag{7.13}$$

$$y_{n+1} = \frac{y_{n-1}}{p_n + q_n y_n + y_{n-1}}, \quad n = 0, 1, \ldots . \tag{7.14}$$

Open Problem 7.5.5 *Assume that* $\{p_n\}_{n=0}^{\infty}$ *and* $\{q_n\}_{n=0}^{\infty}$ *are period-two sequences of nonnegative real numbers. Investigate the global character of all positive solutions of Eqs(7.12)-(7.14). Extend and generalize.*

Chapter 8

$(3, 1)$-Type Equations

8.1 Introduction

Eq(1) contains the following three equations of the $(1, 3)$-type:

$$x_{n+1} = \frac{\alpha + \beta x_n + \gamma x_{n-1}}{A}, \quad n = 0, 1, \ldots \tag{8.1}$$

$$x_{n+1} = \frac{\alpha + \beta x_n + \gamma x_{n-1}}{B x_n}, \quad n = 0, 1, \ldots \tag{8.2}$$

and

$$x_{n+1} = \frac{\alpha + \beta x_n + \gamma x_{n-1}}{C x_{n-1}}, \quad n = 0, 1, \ldots . \tag{8.3}$$

Please recall our classification convention in which all parameters that appear in these equations are positive, the initial conditions are nonnegative, and the denominators are always positive.

Eq(8.1) is a linear difference equation and can be solved explicitly.

Eq(8.2) by the change of the variables

$$x_n = y_n + \frac{\beta}{B},$$

is reduced to a $(2, 2)$-type equation of the form of Eq(6.4) (see Section 6.5), namely,

$$y_{n+1} = \frac{(\alpha + \frac{\beta\gamma}{B}) + \gamma y_{n-1}}{\beta + B y_n}, \quad n = 0, 1, \ldots .$$

Eq(8.3) by the change of the variables

$$x_n = y_n + \frac{\gamma}{C},$$

is reduced to a $(2,2)$-type equation of the form of Eq(6.2) (see Section 6.3) namely,

$$y_{n+1} = \frac{(\alpha + \frac{\beta\gamma}{C}) + \beta y_n}{\gamma + C y_{n-1}}, \quad n = 0, 1, \ldots .$$

Hence there is nothing else remaining to do about these equations other than to pose some **Open Problems and Conjectures.**

8.2 Open Problems and Conjectures

Open Problem 8.2.1 *Assume $\gamma > \beta$. Determine the set of initial conditions $(x_{-1}, x_0) \in (0,\infty) \times (0,\infty)$ through which the solutions of Eq(8.2) are bounded.*

Open Problem 8.2.2 *(a) Assume $\gamma = \beta$ and let*

$$\ldots, \phi, \psi, \phi, \psi, \ldots \tag{8.4}$$

be a period-two solution of Eq(8.2). Determine the basin of attraction of (8.4).

(b) Let $\{x_n\}_{n=-1}^{\infty}$ be a positive solution of Eq(8.2) which converges to (8.4). Determine the values of ϕ and ψ in terms of the initial conditions x_{-1} and x_0.

Conjecture 8.2.1 *Show that every positive solution of Eq(8.3) converges to the positive equilibrium of the equation.*

Open Problem 8.2.3 *Determine the set $\mathcal{G}$ of all initial points $(x_{-1}, x_0) \in R \times R$ through which the equation*

$$x_{n+1} = \frac{1 + x_n + x_{n-1}}{x_n}$$

is well defined for all n and for these initial points determine the long-term behavior of the solutions $\{x_n\}_{n=-1}^{\infty}$.

Open Problem 8.2.4 *Determine the set $\mathcal{G}$ of all initial points $(x_{-1}, x_0) \in R \times R$ through which the equation*

$$x_{n+1} = \frac{1 + x_n + x_{n-1}}{x_{n-1}}$$

is well defined for all n and for these initial points determine the long-term behavior of the solutions $\{x_n\}_{n=-1}^{\infty}$.

Open Problem 8.2.5 *Assume that* $\{p_n\}_{n=0}^{\infty}$ *and* $\{q_n\}_{n=0}^{\infty}$ *are convergent sequences of nonnegative real numbers with finite limits,*

$$p = \lim_{n\to\infty} p_n \quad and \quad q = \lim_{n\to\infty} q_n.$$

Investigate the asymptotic behavior and the periodic nature of all positive solutions of each of the following two difference equations:

$$y_{n+1} = \frac{p_n + x_n + x_{n-1}}{q_n x_n}, \quad n = 0, 1, \ldots \tag{8.5}$$

$$y_{n+1} = \frac{p_n + x_n + x_{n-1}}{q_n x_{n-1}}, \quad n = 0, 1, \ldots . \tag{8.6}$$

Open Problem 8.2.6 *Assume that* $\{p_n\}_{n=0}^{\infty}$ *and* $\{q_n\}_{n=0}^{\infty}$ *are period-two sequences of nonnegative real numbers. Investigate the global character of all positive solutions of Eqs(8.5) and (8.6). Extend and generalize.*

Chapter 9

(2, 3)-Type Equations

9.1 Introduction

Eq(1) contains the following three equations of the (2, 3)-type:

$$x_{n+1} = \frac{\alpha + \beta x_n}{A + Bx_n + Cx_{n-1}}, \quad n = 0, 1, \ldots \tag{9.1}$$

$$x_{n+1} = \frac{\alpha + \gamma x_{n-1}}{A + Bx_n + Cx_{n-1}}, \quad n = 0, 1, \ldots \tag{9.2}$$

and

$$x_{n+1} = \frac{\beta x_n + \gamma x_{n-1}}{A + Bx_n + Cx_{n-1}}, \quad n = 0, 1, \ldots . \tag{9.3}$$

Please recall our classification convention in which all the parameters in the above (2, 3)-type equations are positive and the initial conditions are nonnegative.

9.2 The Case $\gamma = 0$: $x_{n+1} = \frac{\alpha+\beta x_n}{A+Bx_n+Cx_{n-1}}$

This is the (2, 3)-type Eq(9.1) which by the change of variables

$$x_n = \frac{A}{B} y_n$$

reduces to the difference equation

$$y_{n+1} = \frac{p + qy_n}{1 + y_n + ry_{n-1}}, \quad n = 0, 1, \ldots \tag{9.4}$$

where

$$p = \frac{\alpha B}{A^2}, \quad q = \frac{\beta}{A}, \quad \text{and} \quad r = \frac{C}{B}.$$

(Eq(9.4) was investigated in [51].)

Here we make some simple observations about linearized stability, invariant intervals, and the convergence of solutions in certain regions of initial conditions.

Eq(9.4) has a unique equilibrium $\overline{y}$ which is positive and is given by

$$\overline{y} = \frac{q - 1 + \sqrt{(q-1)^2 + 4p(r+1)}}{2(r+1)}.$$

By employing the linearized stability Theorem 1.1.1, we obtain the following result:

Theorem 9.2.1 *The equilibrium $\overline{y}$ of Eq(9.4) is locally asymptotically stable for all values of the parameters p, q and r.*

9.2.1 Boundedness of Solutions

We will prove that all solutions of Eq(9.4) are bounded.

Theorem 9.2.2 *Every solution of Eq(9.4) is bounded from above and from below by positive constants.*

Proof. Let $\{y_n\}$ be a solution of Eq(9.4). Clearly, if the solution is bounded from above by a constant M, then

$$y_{n+1} \geq \frac{p}{1 + (1+r)M}$$

and so it is also bounded from below. Now assume for the sake of contradiction that the solution is not bounded from above. Then there exists a subsequence $\{y_{1+n_k}\}_{k=0}^{\infty}$ such that

$$\lim_{k\to\infty} n_k = \infty, \quad \lim_{k\to\infty} y_{1+n_k} = \infty, \quad \text{and} \quad y_{1+n_k} = \max\{y_n : n \leq n_k\} \quad \text{for } k \geq 0.$$

From (9.4) we see that

$$y_{n+1} < q y_n + p \quad \text{for} \quad n \geq 0$$

and so

$$\lim_{k\to\infty} y_{n_k} = \lim_{k\to\infty} y_{n_k - 1} = \infty.$$

Hence for sufficiently large k,

$$0 < y_{1+n_k} - y_{n_k} = \frac{p + [(q-1) - y_{n_k} - r y_{n_k-1}]y_{n_k}}{1 + y_{n_k} + r y_{n_k-1}} < 0$$

which is a contradiction and the proof is complete. □

The above proof has an advantage that extends to several equations of the form of Eq(9.1) with nonnegative parameters. The boundedness of solutions of the special equation (9.4) with positive parameters immediately follows from the observation that if

$$M = \max\{1, p, q\}$$

then

$$y_{n+1} \leq \frac{M + My_n}{1 + y_n} = M \quad for \ n \geq 0.$$

9.2.2 Invariant Intervals

The following result, which can be established by direct calculation, gives a list of invariant intervals for Eq(9.4). Recall that I is an invariant interval, if whenever,

$$y_N, y_{N+1} \in I \quad \text{for some integer} \quad N \geq 0,$$

then

$$y_n \in I \quad \text{for} \quad \text{n} \geq N.$$

Lemma 9.2.1 *Eq(9.4) possesses the following invariant intervals:*

(a)

$$\left[0, \frac{q - 1 + \sqrt{(q-1)^2 + 4p}}{2}\right], \quad when \quad p \leq q;$$

(b)

$$\left[\frac{p-q}{qr}, q\right], \quad when \quad q < p < q(rq+1);$$

(c)

$$\left[q, \frac{p-q}{qr}\right], \quad when \quad p > q(rq+1).$$

Proof.

(a) Set

$$g(x) = \frac{p + qx}{1 + x} \quad \text{and} \quad b = \frac{q - 1 + \sqrt{(q-1)^2 + 4p}}{2}$$

and observe that g is an increasing function and $g(b) \leq b$. Using Eq(9.4) we see that when $y_{k-1}, y_k \in [0, b]$, then

$$y_{k+1} = \frac{p + qy_k}{1 + y_k + ry_{k-1}} \leq \frac{p + qy_k}{1 + y_k} = g(y_k) \leq g(b) \leq b.$$

The proof follows by induction.

(b) It is clear that the function

$$f(x,y) = \frac{p+qx}{1+x+ry}$$

is increasing in x for $y > \frac{p-q}{qr}$. Using Eq(9.4) we see that when $y_{k-1}, y_k \in \left[\frac{p-q}{qr}, q\right]$, then

$$y_{k+1} = \frac{p+qy_k}{1+y_k+ry_{k-1}} = f(y_k, y_{k-1}) \leq f\left(q, \frac{p-q}{qr}\right) = q.$$

Also by using the condition $p < q(rq+1)$ we obtain

$$y_{k+1} = \frac{p+qy_k}{1+y_k+ry_{k-1}} = f(y_k, y_{k-1}) \geq f\left(\frac{p-q}{qr}, q\right) = \frac{q(pr+p-q)}{(rq)^2+rq+p-q} > \frac{p-q}{qr}.$$

The proof follows by the induction.

(c) It is clear that the function

$$f(x,y) = \frac{p+qx}{1+x+ry}$$

is decreasing in x for $y < \frac{p-q}{qr}$. Using Eq(9.4) we see that when $y_{k-1}, y_k \in \left[q, \frac{p-q}{qr}\right]$, then

$$y_{k+1} = \frac{p+qy_k}{1+y_k+ry_{k-1}} = f(y_k, y_{k-1}) \geq f\left(\frac{p-q}{qr}, \frac{p-q}{qr}\right) = q.$$

Also by using the condition $p > q(rq+1)$ we obtain

$$y_{k+1} = \frac{p+qy_k}{1+y_k+ry_{k-1}} = f(y_k, y_{k-1}) \leq f(q,q) = \frac{p+q^2}{1+(r+1)q} < \frac{p-q}{qr}.$$

The proof follows by induction.

□

9.2.3 Semicycle Analysis

Let $\{y_n\}_{n=-1}^{\infty}$ be a solution of Eq(9.4). Then the following identitites are true for $n \geq 0$:

$$y_{n+1} - q = \frac{qr(\frac{p-q}{qr} - y_{n-1})}{1+y_n+ry_{n-1}}, \tag{9.5}$$

$$y_{n+1} - \frac{p-q}{qr} = \frac{qr\left(q - \frac{p-q}{qr}\right)y_n + qr\left(y_{n-1} - \frac{p-q}{qr}\right) + pr(q - y_{n-1})}{qr(1+y_n+ry_{n-1})}, \tag{9.6}$$

$$y_n - y_{n+4} =$$

$$\frac{qr\left(y_n - \frac{p-q}{qr}\right) y_{n+1} + (y_n - q)(y_{n+1}y_{n+3} + y_{n+3} + y_{n+1} + ry_ny_{n+3}) + y_n + ry_n^2 - p}{(1 + y_{n+3})(1 + y_{n+1} + ry_n) + r(p + qy_{n+1})}, \tag{9.7}$$

$$y_{n+1} - \overline{y} = \frac{(\overline{y} - q)(\overline{y} - y_n) + \overline{y}r(\overline{y} - y_{n-1})}{1 + y_n + ry_{n-1}}. \tag{9.8}$$

The proofs of the following two lemmas are straightforward consequences of the Identities (9.5)- (9.8) and will be omitted.

Lemma 9.2.2 *Assume*

$$p > q(qr + 1)$$

and let $\{y_n\}_{n=-1}^{\infty}$ *be a solution of Eq(9.4). Then the following statements are true:*

(i) If for some $N \geq 0, \quad y_N > \frac{p-q}{qr}, \quad$ *then* $\quad y_{N+2} < q;$

(ii) If for some $N \geq 0, \quad y_N = \frac{p-q}{qr}, \quad$ *then* $\quad y_{N+2} = q;$

(iii) If for some $N \geq 0, \quad y_N < \frac{p-q}{qr}, \quad$ *then* $\quad y_{N+2} > q;$

(iv) If for some $N \geq 0, \quad q < y_N < \frac{p-q}{qr}, \quad$ *then* $\quad q < y_{N+2} < \frac{p-q}{qr};$

(v) If for some $N \geq 0, \quad \overline{y} \geq y_{N-1}$ *and* $\overline{y} \geq y_N$*, then* $\quad y_{N+1} \geq \overline{y};$

(vi) If for some $N \geq 0, \quad \overline{y} < y_{N-1}$ *and* $\overline{y} < y_N$*, then* $\quad y_{N+1} < \overline{y};$

(vii) $q < \overline{y} < \frac{p-q}{qr}.$

Lemma 9.2.3 *Assume*

$$q < p < q(qr + 1)$$

and let $\{y_n\}_{n=-1}^{\infty}$ *be a solution of Eq(9.4). Then the following statements are true:*

(i) If for some $N \geq 0, \quad y_N < \frac{p-q}{qr}, \quad$ *then* $\quad y_{N+2} > q;$

(ii) If for some $N \geq 0, \quad y_N = \frac{p-q}{qr}, \quad$ *then* $\quad y_{N+2} = q;$

(iii) If for some $N \geq 0, \quad y_N > \frac{p-q}{qr}, \quad$ *then* $\quad y_{N+2} < q;$

(iv) If for some $N \geq 0, \quad y_N > \frac{p-q}{qr}$ *and* $y_N < q \quad$ *then* $\quad q > y_{N+2} > \frac{p-q}{qr};$

(v) $\frac{p-q}{qr} < \overline{y} < q.$

Lemma 9.2.4 *Assume*

$$p = q(qr + 1)$$

and let $\{y_n\}_{n=-1}^{\infty}$ *be a solution of Eq(9.4). Then*

$$y_{n+1} - q = \frac{qr}{1 + y_n + ry_{n-1}}(q - y_{n-1}). \tag{9.9}$$

Furthermore when $qr < 1$, *then*

$$\lim_{n\to\infty} y_n = \overline{y}. \tag{9.10}$$

Proof. Identity (9.9) follows by straightforward computation. The limit (9.10) is a consequence of the fact that in this case $qr \in (0, \infty)$ and Eq(9.4) has no period-two solution. □

Here we present a semicycle analysis of the solutions of Eq(9.4) when $p = q(1 + rq)$. In this case Eq(9.4) becomes

$$y_{n+1} = \frac{q + rq^2 + qy_n}{1 + y_n + ry_{n-1}}, \quad n = 0, 1, \ldots$$

and the only equilibrium point is $\overline{y} = q$.

Theorem 9.2.3 *Suppose that* $p = q + rq^2$ *and let* $\{y_n\}_{n=-1}^{\infty}$ *be a nontrivial solution of Eq(9.4). Then this solution is oscillatory and the sum of the lengths of two consecutive semicycles, excluding the first, is equal to four. More precisely, the following statements are true for all* $k \geq 0$:

(i)
$$y_{-1} > q \text{ and } y_0 \leq q \Longrightarrow y_{4k-1} > q, y_{4k} \leq 1, y_{4k+1} < q, \text{ and } y_{4k+2} \geq q;$$

(ii)
$$y_{-1} < q \text{ and } y_0 \geq q \Longrightarrow y_{4k-1} < q, y_{4k} \geq 1, y_{4k+1} > q, \text{ and } y_{4k+2} \leq q;$$

(iii)
$$y_{-1} > q \text{ and } y_0 \geq q \Longrightarrow y_{4k-1} > q, y_{4k} \geq 1, y_{4k+1} < q, \text{ and } y_{4k+2} \leq q;$$

(iv)
$$y_{-1} < q \text{ and } y_0 \leq q \Longrightarrow y_{4k-1} < q, y_{4k} \leq 1, y_{4k+1} > q, \text{ and } y_{4k+2} \geq q.$$

Proof. The proof follows from identity (9.5). □

9.2.4 Global Asymptotic Stability

The following lemma establishes that when $p \neq q(qr+1)$, every solution of Eq(9.4) is eventually trapped into one of the three invariant intervals of Eq(9.4) described in Lemma 9.2.1. More precisely, the following is true:

Lemma 9.2.5 *Let I denote the interval which is defined as follows:*

$$I = \begin{cases} [0, \frac{q-1+\sqrt{(q-1)^2+4p}}{2}] & \text{if} \quad p \leq q; \\ [\frac{p-q}{qr}, q] & \text{if} \quad q < p < q(qr+1); \\ [q, \frac{p-q}{qr}] & \text{if} \quad p > q(qr+1). \end{cases}$$

Then every solution of Eq(9.4) lies eventually in I.

Proof. Let $\{y_n\}_{n=-1}^{\infty}$ be a solution of Eq(9.4). First assume that

$$p \leq q.$$

Then clearly, $\overline{y} \in I$. Set

$$b = \frac{q-1+\sqrt{(q-1)^2+4p}}{2}$$

and observe that

$$y_{n+1} - b = \frac{(q-p)(y_n - b) - r(p+qb)y_{n-1}}{(1+y_n+ry_{n-1})(1+b)}, \quad n \geq 0.$$

Hence, if for some N, $y_N \leq b$, then $y_{N+1} < b$. Now assume for the sake of contradiction that

$$y_n > b \quad \text{for} \quad n > 0.$$

Then $y_n > \overline{y}$ for $n \geq 0$ and so

$$\lim_{n\to\infty} y_n = \overline{y} \in I$$

which is a contradiction.

Next assume that

$$q < p < q(qr+1)$$

and, for the sake of contradiction, also assume that the solution $\{y_n\}_{n=-1}^{\infty}$ is not eventually in the interval I. Then by Lemma 9.2.3, there exists $N > 0$ such that one of the following three cases holds:

(i) $y_N > q, y_{N+1} > q,$ and $y_{N+2} < \frac{p-q}{qr}$;

(ii) $y_N > q, y_{N+1} < \frac{p-q}{qr},$ and $y_{N+2} < \frac{p-q}{qr}$;

(iii) $y_N > q$, $\frac{p-q}{qr} \leq y_{N+1} \leq q$, and $y_{N+2} < \frac{p-q}{qr}$.

Also observe that

if $y_N \geq q$, then $y_N + ry_N^2 - p > 0$

and

if $y_N \leq \frac{p-q}{qr}$, then $y_N + ry_N^2 - p < 0$.

The desired contradiction is now obtained by using the Identity (9.7) from which it follows that for $j \in \{0, 1, 2, 3\}$, each subsequence $\{y_{n+4k+j}\}_{k=0}^{\infty}$ with all its terms outside the interval I converges monotonically and enters in the interval I.

The proof when

$$p > q(qr + 1)$$

is similar and will be omitted. □

By using the monotonic character of the function

$$f(x, y) = \frac{p + qx}{1 + x + ry}$$

in each of the intervals in Lemma 9.2.1, together with the appropriate convergence Theorems 1.4.7 and 1.4.5, we can obtain some global asymptotic stability results for the solutions of Eq(9.4). For example, the following results are true for Eq(9.4):

Theorem 9.2.4 *(a) Assume that either*

$$p \geq q + q^2 r,$$

or

$$p < \frac{q}{1 + r}.$$

Then the equilibrium $\overline{y}$ *of Eq(9.4) is globally asymptotically stable.*

(b) Assume that either

$$p \leq q$$

or

$$q < p < q + q^2 r$$

and that one of the following conditions is also satisfied:

(i) $q \leq 1$;

(ii) $r \leq 1$;

(iii) $r > 1$ *and* $(q-1)^2(r-1) \leq 4p$.

Then the equilibrium $\overline{y}$ *of Eq(9.4) is globally asymptotically stable.*

Proof.

(a) The proof follows from Lemmas 9.2.1 and 9.2.4 and Theorems 1.4.7 and 1.4.2.

(b) In view of Lemma 9.2.1 we see that when $y_{-1}, y_0 \in \left[0, \frac{q-1+\sqrt{(q-1)^2+4p}}{2}\right]$, then $y_n \in \left[0, \frac{q-1+\sqrt{(q-1)^2+4p}}{2}\right]$ for all $n \geq 0$. It is easy to check that $\bar{y} \in \left[0, \frac{q-1+\sqrt{(q-1)^2+4p}}{2}\right]$ and that in the interval $\left[0, \frac{q-1+\sqrt{(q-1)^2+4p}}{2}\right]$, the function f increases in x and decreases in y.

We will employ Theorem 1.4.5 and so it remains to be shown that if

$$m = f(m, M) \quad \text{and} \quad M = f(M, m)$$

then $M = m$. This system has the form

$$m = \frac{p + qm}{1 + m + rM} \quad \text{and} \quad M = \frac{p + qM}{1 + M + rm}.$$

Hence $(M - m)(1 - q + M + m) = 0$. Now if $m + M \neq q - 1$, then $M = m$. For instance, this is the case if Condition (i) is satisfied. If $m + M = q - 1$ then m and M satisfy the equation

$$(r - 1)m^2 + (r - 1)(1 - q)m + p = 0.$$

Clearly now if Condition (ii) or (iii) is satisfied, $m = M$ from which the result follows.

The proof when $q < p < q + q^2 r$ holds follows from Lemma 9.2.1 and Theorem 1.4.5.

□

9.3 The Case $\beta = 0: \ x_{n+1} = \frac{\alpha+\gamma x_{n-1}}{A+Bx_n+Cx_{n-1}}$

This is the $(2,3)$-type Eq(9.2) which by the change of variables

$$x_n = \frac{A}{B} y_n$$

reduces to the difference equation

$$y_{n+1} = \frac{p + q y_{n-1}}{1 + y_n + r y_{n-1}}, \quad n = 0, 1, \ldots \tag{9.11}$$

where

$$p = \frac{\alpha B}{A^2}, \quad q = \frac{\gamma}{A}, \quad \text{and} \quad r = \frac{C}{B}.$$

This equation has not been investigated yet. Here we make some simple observations about linearized stability, invariant intervals, and the convergence of solutions in certain regions of initial conditions.

Eq(9.11) has a unique equilibrium $\overline{y}$ which is positive and is given by

$$\overline{y} = \frac{q-1+\sqrt{(q-1)^2+4p(r+1)}}{2(r+1)}.$$

By employing the linearized stability Theorem 1.1.1 we obtain the following result.

Theorem 9.3.1 *(a) The equilibrium $\overline{y}$ of Eq(9.11) is locally asymptotically stable when either*

$$q \leq 1$$

or

$$q > 1 \quad and \quad (r-1)(q-1)^2 + 4pr^2 > 0.$$

(b) The equilibrium $\overline{y}$ of Eq(9.11) is unstable, and more precisely a saddle point, when

$$q > 1 \quad and \quad (r-1)(q-1)^2 + 4pr^2 < 0. \tag{9.12}$$

9.3.1 Existence and Local Stability of Period-Two Cycles

Let

$$\ldots, \phi, \psi, \phi, \psi, \ldots$$

be a period-two cycle of Eq(9.11). Then

$$\phi = \frac{p+q\phi}{1+\psi+r\phi} \quad \text{and} \quad \psi = \frac{p+q\psi}{1+\phi+r\psi}.$$

It now follows after some calculation that the following result is true.

Lemma 9.3.1 *Eq(9.11) has a prime period-two solution*

$$\ldots, \phi, \psi, \phi, \psi, \ldots$$

if and only if (9.12) holds.

Furthermore when (9.12) holds, the period-two solution is "unique" and the values of ϕ and ψ are the positive roots of the quadratic equation

$$t^2 - \frac{q-1}{r}t + \frac{p}{1-r} = 0.$$

To investigate the local stability of the two cycle

$$\ldots, \phi, \psi, \phi, \psi, \ldots$$

we set

$$u_n = y_{n-1} \quad \text{and} \quad v_n = y_n, \quad \text{for } n = 0, 1, \ldots$$

and write Eq(9.11) in the equivalent form:

$$u_{n+1} = v_n$$

$$v_{n+1} = \frac{p+qu_n}{1+v_n+ru_n}, \quad n = 0, 1, \ldots .$$

Let T be the function on $(0, \infty) \times (0, \infty)$ defined by:

$$T\begin{pmatrix} u \\ v \end{pmatrix} = \begin{pmatrix} v \\ \frac{p+qu}{1+v+ru} \end{pmatrix}.$$

Then

$$\begin{pmatrix} \phi \\ \psi \end{pmatrix}$$

is a fixed point of T^2, the second iterate of T. By a simple calculation we find that

$$T^2\begin{pmatrix} u \\ v \end{pmatrix} = \begin{pmatrix} g(u,v) \\ h(u,v) \end{pmatrix}$$

where

$$g(u,v) = \frac{p+qu}{1+v+ru} \quad \text{and} \quad h(u,v) = \frac{p+qv}{1+g(u,v)+rv}.$$

Clearly the two cycle is locally asymptotically stable when the eigenvalues of the Jacobian matrix J_{T^2}, evaluated at $\begin{pmatrix} \phi \\ \psi \end{pmatrix}$, lie inside the unit disk.

We have

$$J_{T^2}\begin{pmatrix} \phi \\ \psi \end{pmatrix} = \begin{pmatrix} \frac{\partial g}{\partial u}(\phi,\psi) & \frac{\partial g}{\partial v}(\phi,\psi) \\ \frac{\partial h}{\partial u}(\phi,\psi) & \frac{\partial h}{\partial v}(\phi,\psi) \end{pmatrix},$$

where

$$\begin{aligned} \frac{\partial g}{\partial u}(\phi,\psi) &= \frac{q-pr+q\psi}{(1+\psi+r\phi)^2}, \\ \frac{\partial g}{\partial v}(\phi,\psi) &= \frac{-p-q\phi}{(1+\psi+r\phi)^2}, \end{aligned}$$

$$\frac{\partial h}{\partial u}(\phi,\psi) = \frac{(p+q\psi)(pr-q-q\psi)}{(1+\psi+r\phi)^2(1+\phi+r\psi)^2},$$
$$\frac{\partial h}{\partial v}(\phi,\psi) = \frac{(p+q\psi)(p+q\phi)}{(1+\psi+r\phi)^2(1+\phi+r\psi)^2}+\frac{q-pr+q\phi}{(1+\phi+r\psi)^2}.$$

Set

$$\mathcal{S}=\frac{\partial g}{\partial u}(\phi,\psi)+\frac{\partial h}{\partial v}(\phi,\psi)$$

and

$$\mathcal{D}=\frac{\partial g}{\partial u}(\phi,\psi)\frac{\partial h}{\partial v}(\phi,\psi)-\frac{\partial g}{\partial v}(\phi,\psi)\frac{\partial h}{\partial u}(\phi,\psi).$$

Then it follows from Theorem 1.1.1 that both eigenvalues of $J_{T^2}\begin{pmatrix}\phi\\ \psi\end{pmatrix}$ lie inside the unit disk $|\lambda|<1$, if and only if

$$|\mathcal{S}|<1+\mathcal{D}<2. \tag{9.13}$$

Inequality (9.13) is equivalent to the following three inequalities:

$$\mathcal{D}<1 \tag{9.14}$$

$$\mathcal{S}<1+\mathcal{D} \tag{9.15}$$

$$-1-\mathcal{D}<\mathcal{S}. \tag{9.16}$$

First we will establish Inequality (9.14). This inequality is equivalent to

$$(q-pr+q\phi)(q-pr+q\psi)-(1+\psi+r\phi)^2(1+\phi+r\psi)^2<0.$$

Observe that,

$$(q-pr+q\phi)(q-pr+q\psi)=(q-pr)^2+q(q-pr)(\phi+\psi)+q^2\phi\psi=$$
$$=\frac{-2q^2r+q^2r^2+3qpr^2-2qpr^3-p^2r^3+p^2r^4-q^3+rq^3+q^2-pr^2q^2-qpr}{r(-1+r)}$$

and that

$$(1+\psi+r\phi)^2(1+\phi+r\psi)^2$$

$$=\left(1+\phi+r\psi+\psi+\psi\phi+r\psi^2+r\phi+\phi^2r+r^2\phi\psi\right)^2$$

$$=\left(\frac{-pr^2+qr+pr-q+q^2}{r}\right)^2.$$

Thus

$$(q - pr + q\phi)(q - pr + q\psi) - (1 + \psi + r\phi)^2 (1 + \phi + r\psi)^2$$

$$= \frac{-pr^3q^2 + 2q^2r - q^2r^2 - q^2 + 3p^2r^3 - 2p^2r^4 - 3rq^3 + 3qpr^3 - 5qpr^2}{r^2(1-r)}$$

$$+ \frac{2qpr + 4pr^2q^2 + 2q^3 + q^4r - p^2r^2 - 2prq^2 - q^4 + r^2q^3}{r^2(1-r)}$$

$$= \frac{(2r^2p - rp + q - q^2 - qr + q^2r)(-r^2p + rp - q + q^2 + qr)}{r^2(1-r)}.$$

Note that

$$-r^2p + rp - q + q^2 + qr = rp(1-r) + q(q-1) + qr > 0.$$

Also by using Condition (9.12) we see that

$$(r-1)(q-1)^2 + 4pr^2 = q^2r - 2qr + r - q^2 + 2q - 1 + 4r^2p < 0$$

and so

$$2r^2p - rp + q - q^2 - qr + q^2r < 2r^2p - rp + q - qr + 2qr - r - 2q + 1 - 4r^2p$$

$$= -2r^2p - rp - q + qr - r + 1 = -2r^2p - rp + (r-1)(q-1) < 0$$

from which the result follows.

Next we turn to Inequality (9.15). This inequality is equivalent to

$$(q - pr + q\psi)(1 + \phi + r\psi)^2 + (p + q\psi)(p + q\phi) + (q - pr + q\phi)(1 + \psi + r\phi)^2$$
$$- (q - pr + q\phi)(q - pr + q\psi) - (1 + \psi + r\phi)^2(1 + \phi + r\psi)^2 < 0.$$

After some lengthy calculation one can see that this inequality is a consequence of Condition (9.12).

Finally Inequality (9.16) is equivalent to

$$(q - pr + q\psi)(1 + \phi + r\psi)^2 + (p + q\psi)(p + q\phi) + (q - pr + q\phi)(1 + \psi + r\phi)^2$$
$$+ (q - pr + q\phi)(q - pr + q\psi) + (1 + \psi + r\phi)^2(1 + \phi + r\psi)^2 > 0.$$

After some lengthy calculation one can see that this inequality is also a consequence of Condition (9.12).

In summary, we have established the following result:

Theorem 9.3.2 *Eq(9.11) has a unique prime period-two solution*

$$\ldots, \phi, \psi, \phi, \psi, \ldots$$

if and only if (9.12) holds.

Furthermore when (9.12) holds, this period-two solution is locally asymptotically stable.

9.3.2 Invariant Intervals

The following result, which can be established by direct calculation, gives a list of invariant intervals for Eq(9.11).

Lemma 9.3.2 *Eq(9.11) possesses the following invariant intervals:*

(a)

$$\left[0, \frac{q-1+\sqrt{(q-1)^2+4pr}}{2r}\right] \quad \textit{when} \quad pr \leq q;$$

(b)

$$\left[\frac{pr-q}{q}, \frac{q}{r}\right] \quad \textit{when} \quad 1 < q < pr < q + \frac{q^2}{r} - \frac{q}{r};$$

(c)

$$\left[\frac{q}{r}, \frac{pr-q}{q}\right] \quad \textit{when} \quad pr \geq q + \frac{q^2}{r}.$$

Proof.

(a) Set

$$g(x) = \frac{p+qx}{1+rx} \quad \text{and} \quad b = \frac{q-1+\sqrt{(q-1)^2+4pr}}{2r}$$

and observe that g is an increasing function and $g(b) \leq b$. Using Eq(9.11) we see that when $y_{k-1}, y_k \in [0, b]$, then

$$y_{k+1} = \frac{p+qy_{k-1}}{1+y_k+qy_{k-1}} \leq \frac{p+qy_{k-1}}{1+ry_{k-1}} = g(y_{k-1}) \leq g(b) \leq b.$$

The proof follows by induction.

(b) It is clear that the function

$$f(x,y) = \frac{p+qy}{1+x+ry}$$

is increasing in y for $x > \frac{pr-q}{q}$. Using Eq(9.11) we see that when $y_{k-1}, y_k \in \left[\frac{pr-q}{q}, \frac{q}{r}\right]$, then

$$y_{k+1} = \frac{p+qy_{k-1}}{1+y_k+ry_{k-1}} = f(y_k, y_{k-1}) \geq f\left(\frac{q}{r}, \frac{pr-q}{q}\right) \geq \frac{pr-q}{q}.$$

Now by using the monotonic character of the function $f(x,y)$ and the condition $q < pr < q + \frac{q^2}{r}$ we obtain

$$y_{k+1} = \frac{p+qy_{k-1}}{1+y_k+ry_{k-1}} = f(y_k, y_{k-1}) \leq f\left(\frac{pr-q}{q}, \frac{q}{r}\right) = \frac{q}{r}.$$

The proof follows by induction.

(c) It is clear that the function

$$f(x,y) = \frac{p+qy}{1+x+ry}$$

is decreasing in y for $x < \frac{pr-q}{q}$. Using Eq(9.11) we see that when $y_{k-1}, y_k \in \left[\frac{q}{r}, \frac{pr-q}{q}\right]$, then

$$y_{k+1} = \frac{p+qy_{k-1}}{1+y_k+qy_{k-1}} = f(y_k, y_{k-1}) \geq f\left(\frac{pr-q}{q}, \frac{pr-q}{q}\right) = \frac{q}{r}.$$

By using the fact that $f(x,y)$ is decreasing in both arguments and the condition $pr > q + \frac{q}{r^2}$ we obtain

$$y_{k+1} = \frac{p+qy_{k-1}}{1+y_k+qy_{k-1}} = f(y_k, y_{k-1}) \leq f\left(\frac{q}{r}, \frac{q}{r}\right)$$

$$= \frac{pr+q^2}{r+(r+1)q} < \frac{pr-q}{q}.$$

The proof follows by induction.

□

9.3.3 Convergence of Solutions

By using the monotonic character of the function

$$f(x,y) = \frac{p+qy}{1+x+ry}$$

in each of the intervals in Lemma 9.3.2, together with the appropriate convergence theorem (from among Theorems 1.4.5-1.4.8) we can obtain some convergence results for the solutions with initial conditions in the invariant intervals. For example, the following results are true for Eq(9.11):

Theorem 9.3.3 *Assume that*

$$pr \le q$$

and that one of the following conditions is also satisfied:

(i) $q \le 1$;

(ii) $r \ge 1$;

(iii) $r < 1$ *and* $(r-1)(q-1)^2 + 4pr^2 \ge 0.$

Then every solution of Eq(9.11) with initial conditions in the invariant interval

$$\left[0, \frac{q-1+\sqrt{(q-1)^2+4pr}}{2r}\right]$$

converges to the equilibrium $\overline{y}$.

Proof. In view of Lemma 9.3.2 we see that when $y_{-1}, y_0 \in \left[0, \frac{q-1+\sqrt{(q-1)^2+4pr}}{2r}\right]$, then $y_n \in \left[0, \frac{q-1+\sqrt{(q-1)^2+4pr}}{2r}\right]$ for all $n \ge 0$. It is easy to check that $\overline{y} \in \left[0, \frac{q-1+\sqrt{(q-1)^2+4pr}}{2r}\right]$ and that in the interval $\left[0, \frac{q-1+\sqrt{(q-1)^2+4pr}}{2r}\right]$ the function f decreases in x and increases in y. The result now follows by employing Theorem 1.4.6.

$\square$

Theorem 9.3.4 *(a) Assume that* $1 < q < pr < q + \frac{q^2}{r} - \frac{q}{r}$ *and that one of the following conditions is also satisfied:*

(i) $r \ge 1$;

(ii) $r < 1$ *and* $(r-1)(q-1)^2 + 4pr^2 \ge 0.$

Then every solution of Eq(9.11) with initial conditions in the invariant interval

$$\left[\frac{pr-q}{q}, \frac{q}{r}\right]$$

converges to the equilibrium $\overline{y}$.

(b) *Assume that* $pr > q + \frac{q^2}{r}$. *Then every solution of Eq(9.11) with initial conditions in the invariant interval*

$$\left[\frac{q}{r}, \frac{pr-q}{q}\right]$$

converges to the equilibrium $\overline{y}$.

Proof.

(a) In view of Lemma 9.3.2 we conclude that $y_n \in \left[\frac{pr-q}{q}, \frac{q}{r}\right]$ for all $n \geq 0$. It is easy to check that $\overline{y} \in \left[\frac{pr-q}{q}, \frac{q}{r}\right]$ and that the function $f(x, y)$ decreases in x and increases in y. The result now follows by employing Theorem 1.4.6.

(b) In view of Lemma 9.3.2 we conclude that $y_n \in \left[\frac{q}{r}, \frac{pr-q}{q}\right]$ for all $n \geq 0$. It is easy to check that $\overline{y} \in \left[\frac{q}{r}, \frac{pr-q}{q}\right]$ and that in the interval $\left[\frac{q}{r}, \frac{pr-q}{q}\right]$ the function f decreases in both x and in y. The result now follows by employing Theorem 1.4.7.

□

9.3.4 Global Stability

By applying Theorem 1.4.3 to Eq(9.11), we obtain the following global asymptotic stability result.

Theorem 9.3.5 *Assume*

$$r \geq 1 \quad \text{and} \quad pr \leq q.$$

Then the positive equilibrium of Eq(9.11) is globally asymptotically stable.

9.3.5 Semicycle Analysis When $pr = q + \frac{q^2}{r}$

Here we present a semicycle analysis of the solutions of Eq(9.11) when $pr = q + \frac{q^2}{r}$. In this case Eq(9.11) becomes

$$y_{n+1} = \frac{qr + q^2 + qr^2 y_{n-1}}{r^2 + r^2 y_n + r^3 y_{n-1}}, \quad n = 0, 1, \ldots$$

and the only positive equilibrium is $\overline{y} = \frac{q}{r}$.

Theorem 9.3.6 *Suppose that* $pr = q + \frac{q^2}{r}$ *and let* $\{y_n\}_{n=-1}^{\infty}$ *be a nontrivial solution of Eq(9.11). Then after the first semicycle, the solution oscillates about the equilibrium* $\overline{y}$ *with semicycles of length one.*

Proof. The proof follows from the relation

$$y_{n+1} - \frac{q}{r} = (\frac{q}{r} - y_n)\frac{qr^2}{r^3(1 + y_n + ry_{n-1})}.$$

□

9.4 The Case $\alpha = 0$: $x_{n+1} = \frac{\beta x_n + \gamma x_{n-1}}{A + Bx_n + Cx_{n-1}}$

This is the (2, 3)-type Eq(9.3) which by the change of variables

$$x_n = \frac{\gamma}{C} y_n,$$

reduces to the difference equation

$$y_{n+1} = \frac{py_n + y_{n-1}}{r + qy_n + y_{n-1}}, \quad n = 0, 1, \ldots \tag{9.17}$$

where

$$p = \frac{\beta}{\gamma}, \quad q = \frac{B}{C}, \quad \text{and } r = \frac{A}{\gamma}.$$

This equation has not been investigated yet. Here we make some simple observations about linearized stability, invariant intervals, and the convergence of solutions in certain regions of initial conditions.

The equilibrium points of Eq(9.17) are the nonnegative solutions of the equation

$$\overline{y} = \frac{(p+1)\overline{y}}{r + (1+q)\overline{y}}.$$

Hence zero is always an equilibrium point and when

$$p + 1 > r, \tag{9.18}$$

Eq(9.17) also possesses the unique positive equilibrium

$$\overline{y} = \frac{p + 1 - r}{q + 1}.$$

The following theorem is a consequence of Theorem 1.1.1 and Theorem 1.3.1.

Theorem 9.4.1 *(a) Assume that*

$$p + 1 \leq r.$$

Then the zero equilibrium of Eq(9.17) is globally asymptotically stable.

(b) Assume that

$$p + 1 > r.$$

Then the zero equilibrium of Eq(9.17) is unstable. Furthermore the zero equilibrium is a saddle point when

$$1 - p < r < 1 + p$$

and a repeller when

$$r < 1 - p.$$

The linearized equation associated with Eq(9.17) about its positive equilibrium $\overline{y}$ is

$$z_{n+1} - \frac{p - q + qr}{(p+1)(q+1)} z_n - \frac{q - p + r}{(p+1)(q+1)} z_{n-1} = 0, \quad n = 0, 1, \ldots .$$

The following result is a consequence of Theorem 1.1.1.

Theorem 9.4.2 *Assume that (9.18) holds. Then the positive equilibrium of Eq(9.17) is locally asymptotically stable when*

$$q + r < 3p + 1 + qr + pq, \tag{9.19}$$

and unstable (a saddle point) when

$$q + r > 3p + 1 + qr + pq. \tag{9.20}$$

Concerning prime period-two solutions for Eq(9.17), it follows from Section 2.5 that the following result is true.

Theorem 9.4.3 *Eq(9.17) has a prime period-two solution*

$$\ldots \phi, \psi, \phi, \psi, \ldots$$

if and only if (9.20) holds. Furthermore when (9.20) holds there is a unique period-two solution and the values of ϕ and ψ are the positive roots of the quadratic equation

$$t^2 - (1 - p - r)t + \frac{p(1 - p - r)}{q - 1} = 0.$$

9.4.1 Invariant Intervals

The following result, which can be established by direct calculation, gives a list of invariant intervals for Eq(9.17).

Lemma 9.4.1 *Eq(9.17) possesses the following invariant intervals:*

(a)

$$\left[0, \frac{p+q-r}{q+1}\right] \quad \textit{when} \quad p = q^2 \quad \textit{and} \quad q^2 + q > r$$

and

$$[0, K] \quad \textit{when} \quad p = q^2 \quad \textit{and} \quad q^2 + q \leq r,$$

where $K > 0$ is an arbitrary number;

(b)

$$\left[0, \frac{qr}{p-q^2}\right] \quad \textit{when} \quad q^2 < p \leq q^2 + r \quad \textit{and} \quad p + q > r,$$

$$[0, K] \quad \textit{when} \quad q^2 < p \leq q^2 + r \quad \textit{and} \quad p + q \leq r,$$

where $K > 0$ is an arbitrary number

and

$$\left[\frac{qr}{p-q^2}, K_1\right] \quad \textit{when} \quad p > q^2 + r$$

where

$$K_1 = \frac{(q-r)(p-q^2) - qr + \sqrt{((r-q)(p-q^2)+qr)^2 + 4q^3r(p-q^2)}}{2q(p-q^2)};$$

(c)

$$\left[0, \frac{pr}{q^2-p}\right] \quad \textit{when} \quad q^2 - qr \leq p < q^2 \quad \textit{and} \quad p + q > r,$$

$$[0, K] \quad \textit{when} \quad q^2 - qr \leq p < q^2 \quad \textit{and} \quad p + q \leq r,$$

where $K > 0$ is an arbitrary number

and

$$\left[\frac{pr}{q^2-p}, K_2\right] \quad \textit{when} \quad q^2 > p + qr,$$

where

$$K_2 = \frac{(q-r)(q^2-p) + \sqrt{((q-r)(q^2-p))^2 + 4p^3qr^2}}{2pqr}.$$

Proof.

(a) It is easy to see that when $p = q^2$ the function

$$f(x, y) = \frac{px + qy}{r + qx + y}$$

is increasing in both arguments. Therefore for any $K > 0$,

$$0 \leq f(x, y) \leq f(K, K).$$

Now observe that the inequality $f(K, K) \leq K$ is equivalent to $q^2 + q - r \leq (q+1)K$. When $q^2 + q \leq r$, this inequality is satisfied for any $K > 0$, while when $q^2 + q > r$ the inequality is true when

$$K \geq \frac{p + q - r}{q + 1}.$$

(b) It is clear that the function $f(x, y)$ is increasing in both arguments for $x < \frac{qr}{p - q^2}$, and it is increasing in x and decreasing in y for $x \geq \frac{qr}{p - q^2}$.

First assume that $q^2 < p \leq q^2 + r$.

Hence for $y_{k-1}, y_k \in \left[0, \frac{qr}{p - q^2}\right]$ we obtain

$$y_{k+1} = f(y_k, y_{k-1}) \leq f(K, K) = (p + q)K \frac{1}{r + (q + 1)K} \leq K,$$

where we set $K = \frac{qr}{p - q^2}$. The last inequality is always satisfied when $p + q \leq r$ and when $p + q > r$, it is satisfied under the condition $r + q^2 > p$.

Second assume that $p > q^2 + r$.

Then for $y_{k-1}, y_k \in \left[\frac{qr}{p - q^2}, K_1\right]$ we can see that

$$f\left(\frac{qr}{p - q^2}, K_1\right) \leq y_{k+1} = f(y_k, y_{k-1}) \leq f(K_1, \frac{qr}{p - q^2}) \leq K_1.$$

(c) It is clear that the function $f(x, y)$ is increasing in both arguments for $y < \frac{pr}{q^2 - p}$, and it is increasing in y and decreasing in x for $y \geq \frac{pr}{q^2 - p}$.

First assume that $q^2 - qr \leq p < q^2$.

Then one can see that for $y_{k-1}, y_k \in \left[0, \frac{pr}{q^2 - p}\right]$ we obtain

$$y_{k+1} = f(y_k, y_{k-1}) \leq f(K, K) = (p + q)K \frac{1}{r + (q + 1)K} \leq K,$$

with $K = \frac{pr}{q^2 - p}$.

Second assume that $p < q^2 - qr$. Then for $y_{k-1}, y_k \in \left[\frac{pr}{q^2 - p}, K_2\right]$ one can see that

$$f\left(\frac{pr}{q^2 - p}, K_2\right) \leq y_{k+1} = f(y_k, y_{k-1}) \leq f\left(K_2, \frac{pr}{q^2 - p}\right) \leq K_2.$$

□

9.4.2 Convergence of Solutions

By using the monotonic character of the function

$$f(x, y) = \frac{px + qy}{r + qx + y}$$

in each of the intervals in Lemma 9.4.1, together with the appropriate convergence theorem (from among Theorems 1.4.5-1.4.8), we can obtain some convergence results for the solutions with initial conditions in the invariant intervals. For example, the following results are true for Eq(9.17):

Theorem 9.4.4 *(a) Assume that $p = q^2$ and $q^2 + q > r$. Then every solution of Eq(9.17) with initial conditions in the invariant interval*

$$\left[0, \frac{q^2 + q - r}{q + 1}\right]$$

converges to the equilibrium $\overline{y}$.

Assume that $p = q^2$ and $q^2 + q \leq r$. Then every solution of Eq(9.17) with initial conditions in the interval $[0, K]$, for any $K > 0$, converges to the equilibrium $\overline{y}$.

(b) Assume that $q^2 + r \geq p > q^2$ and $p + q > r$. Then every solution of Eq(9.17) with initial conditions in the invariant interval

$$\left[0, \frac{qr}{p - q^2}\right]$$

converges to the equilibrium $\overline{y}$.

Assume that $q^2 + r \geq p > q^2$ and $p + q \leq r$. Then every solution of Eq(9.17) with initial conditions in the interval $[0, K]$, for any $K > 0$, converges to the equilibrium $\overline{y}$.

(c) Assume that $q^2 - qr \geq p < q^2$ and $p + q > r$. Then every solution of Eq(9.17) with initial conditions in the invariant interval

$$\left[0, \frac{pr}{q^2 - p}\right]$$

converges to the equilibrium $\overline{y}$.

Assume that $q^2 - qr \leq p < q^2$ and $p + q \leq r$. Then every solution of Eq(9.17) with initial conditions in the interval $[0, K]$, for any $K > 0$, converges to the equilibrium $\overline{y}$.

Proof. The proof is an immediate consequence of Lemma 9.4.1 and Theorem 1.4.8. □

Theorem 9.4.5 *Assume that*

$$p > q^2 + r$$

and that one of the following conditions is also satisfied:

(i) $p \leq q + r$;

(ii) $q \geq 1$;

(iii) $q < 1$ *and* $p \leq \frac{3q^2+q-qr+r}{1-q}$.

Then every solution of Eq(9.11) with initial conditions in the invariant interval $\left[\frac{qr}{p-q^2}, K_1\right]$ *converges to the equilibrium* $\overline{y}$.

Proof. The proof is a consequence of Lemma 9.4.1 and Theorem 1.4.5.

□

Theorem 9.4.6 *Assume that* $q^2 > p + qr$ *and that Condition (9.20) is not satisfied. Then every solution of Eq(9.11) with initial conditions in the invariant interval* $\left[\frac{qr}{p-q^2}, K_2\right]$ *converges to the equilibrium* $\overline{y}$.

Proof. The proof is an immediate consequence of Lemma 9.4.1 and Theorem 1.4.6.

□

9.5 Open Problems and Conjectures

Conjecture 9.5.1 *Assume*

$$\alpha, \beta, A, B, C \in (0, \infty).$$

Show that every positive solution of Eq(9.1) has a finite limit.

Conjecture 9.5.2 *Assume*

$$\alpha, \gamma, A, B, C \in (0, \infty).$$

Show that every positive solution of Eq(9.2) is bounded.

Conjecture 9.5.3 *Assume*

$$\beta, \gamma, A, B, C \in (0, \infty).$$

Show that every positive solution of Eq(9.3) is bounded.

Conjecture 9.5.4 *Assume*

$$\alpha, \gamma, A, B, C \in (0, \infty)$$

and suppose that Eq(9.2) has no prime period-two solutions. Show that the positive equilibrium of Eq(9.2) is globally asymptotically stable.

Conjecture 9.5.5 *Assume that* $x_{-1}, x_0 \in [0, \infty)$ *with* $x_{-1} + x_0 > 0$ *and that*

$$\beta, \gamma, A, B, C \in (0, \infty).$$

Suppose that Eq(9.3) has no prime period-two solutions. Show that the positive equilibrium of Eq(9.3) is globally asymptotically stable.

Open Problem 9.5.1 *Determine the set* $\mathcal{G}$ *of all initial points* $(x_{-1}, x_0) \in R \times R$ *through which the equation*

$$x_{n+1} = \frac{x_n + x_{n-1}}{1 + x_n + x_{n-1}}$$

is well defined for all $n \geq 0$*, and for these initial points investigate the long-term behavior of the solutions* $\{x_n\}_{n=-1}^{\infty}$. *Extend and generalize.*

Open Problem 9.5.2 *Determine the set* $\mathcal{G}$ *of all initial points* $(x_{-1}, x_0) \in R \times R$ *through which the equation*

$$x_{n+1} = \frac{1 + x_n}{1 + x_n + x_{n-1}}$$

is well defined for all $n \geq 0$*, and for these initial points investigate the long-term behavior of the solutions* $\{x_n\}_{n=-1}^{\infty}$. *Extend and generalize.*

Open Problem 9.5.3 *Determine the set* $\mathcal{G}$ *of all initial points* $(x_{-1}, x_0) \in R \times R$ *through which the equation*

$$x_{n+1} = \frac{1 + x_{n-1}}{1 + x_n + x_{n-1}}$$

is well defined for all $n \geq 0$*, and for these initial points investigate the long-term behavior of the solutions* $\{x_n\}_{n=-1}^{\infty}$. *Extend and generalize.*

Open Problem 9.5.4 *Assume that* $\{p_n\}_{n=0}^{\infty}$, $\{q_n\}_{n=0}^{\infty}$, *and* $\{r_n\}_{n=0}^{\infty}$ *are convergent sequences of nonnegative real numbers with finite limits,*

$$p = \lim_{n\to\infty} p_n, \quad q = \lim_{n\to\infty} q_n, \quad \text{and} \quad r = \lim_{n\to\infty} r_n.$$

Investigate the asymptotic behavior and the periodic nature of all positive solutions of each of the following three difference equations:

$$y_{n+1} = \frac{p_n + q_n y_n}{1 + y_n + r_n y_{n-1}}, \quad n = 0, 1, \ldots \tag{9.21}$$

$$y_{n+1} = \frac{p_n + q_n y_{n-1}}{1 + y_n + r_n y_{n-1}}, \quad n = 0, 1, \ldots \tag{9.22}$$

$$y_{n+1} = \frac{p_n y_n + y_{n-1}}{r_n + q_n y_n + y_{n-1}}, \quad n = 0, 1, \ldots . \tag{9.23}$$

Open Problem 9.5.5 *Assume that* $\{p_n\}_{n=0}^{\infty}$, $\{q_n\}_{n=0}^{\infty}$, *and* $\{r_n\}_{n=0}^{\infty}$ *are period-two sequences of nonnegative real numbers. Investigate the global character of all positive solutions of Eqs(9.21)-(9.23). Extend and generalize.*

Open Problem 9.5.6 *Assume that (9.12) holds. Investigate the basin of attraction of the two cycle*

$$\ldots, \phi, \psi, \phi, \psi, \ldots$$

of Eq(9.11).

Open Problem 9.5.7 *Assume that (9.20) holds. Investigate the basin of attraction of the two cycle*

$$\ldots, \phi, \psi, \phi, \psi, \ldots$$

of Eq(9.17).

Conjecture 9.5.6 *Assume that (9.20) holds. Show that the period-two solution*

$$\ldots, \phi, \psi, \phi, \psi, \ldots$$

of Eq(9.17) is locally asymptotically stable.

Open Problem 9.5.8 *Assume that*

$$p, q, r \in [0, \infty) \quad and \quad k \in \{2, 3, \ldots\}.$$

Investigate the global behavior of all positive solutions of each of the following difference equations:

$$y_{n+1} = \frac{p + qy_n}{1 + y_n + ry_{n-k}}, \quad n = 0, 1, \ldots \tag{9.24}$$

$$y_{n+1} = \frac{p + qy_{n-k}}{1 + y_n + ry_{n-k}}, \quad n = 0, 1, \ldots \tag{9.25}$$

$$y_{n+1} = \frac{py_n + y_{n-k}}{r + qy_n + y_{n-k}}, \quad n = 0, 1, \ldots. \tag{9.26}$$

Chapter 10

$(3,2)$-Type Equations

10.1 Introduction

Eq(1) contains the following three equations of the $(3,2)$-type:

$$x_{n+1} = \frac{\alpha + \beta x_n + \gamma x_{n-1}}{A + Bx_n}, \quad n = 0, 1, \ldots \tag{10.1}$$

$$x_{n+1} = \frac{\alpha + \beta x_n + \gamma x_{n-1}}{A + Cx_{n-1}}, \quad n = 0, 1, \ldots \tag{10.2}$$

and

$$x_{n+1} = \frac{\alpha + \beta x_n + \gamma x_{n-1}}{Bx_n + Cx_{n-1}}, \quad n = 0, 1, \ldots . \tag{10.3}$$

Please recall our classification convention in which all parameters that appear in these equations are positive, the initial conditions are nonnegative, and the denominators are always positive.

10.2 The Case $C = 0: \ x_{n+1} = \frac{\alpha+\beta x_n+\gamma x_{n-1}}{A+Bx_n}$

This is the $(3,2)$-type Eq(10.1) which by the change of variables

$$x_n = \frac{A}{B} y_n$$

reduces to the difference equation

$$y_{n+1} = \frac{p + qy_n + ry_{n-1}}{1 + y_n}, \quad n = 0, 1, \ldots \tag{10.4}$$

where

$$p = \frac{\alpha B}{A^2}, \quad q = \frac{\beta}{A}, \quad \text{and } r = \frac{\gamma}{A}.$$

(Eq(10.4) was investigated in [29].)

The linearized equation of Eq(10.4) about its unique equilibrium

$$\overline{y} = \frac{q + r - 1 + \sqrt{(q + r - 1)^2 + 4p}}{2}$$

is

$$z_{n+1} - \frac{q - p - r\overline{y}}{(1 + \overline{y})^2} z_n - \frac{r}{1 + \overline{y}} z_{n-1} = 0, \quad n = 0, 1, \ldots .$$

By applying the linearized stability Theorem 1.1.1 we see that $\overline{y}$ is locally asymptotically stable when

$$r < 1 + q$$

and is an unstable (saddle point) equilibrium when

$$r > 1 + q.$$

When

$$r = q + 1$$

Eq(10.4) possesses prime period-two solutions. See Sections 2.5 and 2.7. The main result in this section is that the solutions of Eq(10.4) exhibit the following **trichotomy character**.

Theorem 10.2.1 *(a) Assume that*

$$r = q + 1. \tag{10.5}$$

Then every solution of Eq(10.4) converges to a period-two solution.

(b) Assume that

$$r < q + 1. \tag{10.6}$$

Then the equilibrium of Eq(10.4) is globally asymptotically stable.

(c) Assume that

$$r > q + 1. \tag{10.7}$$

Then Eq(10.4) possesses unbounded solutions.

The proof of part *(a)* of Theorem 10.2.1 was given, for a more general equation, in Section 2.7. The proofs of parts *(b)* and *(c)* of Theorem 10.2.1 are given in Sections 10.2.1 and 10.2.2 below.

10.2.1 Proof of Part *(b)* of Theorem 10.2.1

Here we assume that Eq(10.6) holds and we prove that the equilibrium of Eq(10.4) is globally asymptotically stable. We know already that when Eq(10.6) holds the equilibrium is locally asymptotically stable so it remains to be shown that every solution of Eq(10.4) is attracted to the equilibrium.

Observe that

$$y_{n+1} = q + \frac{(p-q) + ry_{n-1}}{1+y_n}, \quad n = 0, 1, \ldots$$

and so when

$$p \geq q \tag{10.8}$$

we have

$$y_n \geq q \quad \text{for} \quad n \geq 1.$$

Therefore the change of variables

$$y_n = q + z_n$$

transforms Eq(10.4) to the difference equation

$$z_{n+1} = \frac{(p + rq - q) + rz_{n-1}}{(1+q) + z_n}, \quad n = 1, \ldots \tag{10.9}$$

which is of the form of the $(2,2)$-type Eq(6.23) which was investigated in Section 6.5 and therefore the proof is complete when (10.8) holds.

So in the remaining part of the proof we assume that

$$p < q.$$

Now observe that the solutions of Eq(10.4) satisfy the following identity

$$y_{n+1} - q = \frac{r}{1+y_n}\left(y_{n-1} - \frac{q-p}{r}\right) \quad \text{for} \quad n \geq 0. \tag{10.10}$$

Note that when

$$q = \frac{q-p}{r}$$

that is when

$$p + qr - q = 0 \tag{10.11}$$

the Identity (10.10) reduces to

$$y_{n+1} - q = \frac{r}{1+y_n}(y_n - q) \quad \text{for} \quad n \geq 0. \tag{10.12}$$

Also in this case

$$0 < r = 1 - \frac{p}{q} < 1$$

and the equilibrium of Eq(10.4) is equal to q.

It follows from (10.12) that

$$|y_{n+1} - q| < r|y_{n+1} - q|$$

and so

$$\lim_{n\to\infty} y_n = q$$

which completes the proof when (10.11) holds.

Still to be considered are cases where

$$p + qr - q > 0 \tag{10.13}$$

and

$$p + qr - q < 0. \tag{10.14}$$

First assume that (10.13) holds. To complete the proof in this case it is sufficient to show that eventually

$$y_n \geq q \tag{10.15}$$

and then use the known result for Eq(10.9). To this end, note that now

$$\frac{q-p}{r} < q$$

and employ (10.10). If (10.15) is not eventually true then either

$$y_{2n} < \frac{q-p}{r} \quad \text{for} \quad n \geq 0 \tag{10.16}$$

or

$$y_{2n-1} < \frac{q-p}{r} \quad \text{for} \quad n \geq 0. \tag{10.17}$$

This is because when

$$y_{N-1} > \frac{q-p}{r} \quad \text{for some} \quad N \geq 0,$$

then it follows from (10.10) that

$$y_{N+1+2K} > q \quad \text{for all} \quad K \geq 0.$$

We will assume that (10.16) holds. The case where (10.17) holds is similar and will be omitted. Then we can see from (10.10) that

$$y_{2n+2} - q = \frac{r}{1 + y_{n+1}} \left(y_{2n} - \frac{q-p}{r}\right) > r\left(y_{2n} - \frac{q-p}{r}\right)$$

and so

$$y_{2n+2} > ry_{2n} + p \quad \text{for} \quad n \geq 0. \tag{10.18}$$

This is impossible when $r \geq 1$, because by (10.16) the solution is bounded. On the other hand, when $r < 1$, it follows from (10.16) and (10.18) that

$$\frac{q-p}{r} > y_{2n+2} > r^{n+1}y_0 + r^n p + \ldots + rp + p$$

and so

$$\frac{q-p}{r} \geq \frac{p}{1-r}$$

which contradicts (10.13) and completes the proof when (10.13) holds.

Next assume that (10.14) holds. Now we claim that every solution of Eq(10.4) lies eventually in the interval $\left[0, \frac{q-p}{r}\right]$. Once we establish this result, the proof that the equilibrium is a global attractor of all solutions would be a consequence of Theorem 1.4.8 and the fact that in this case the function

$$f(x,y) = \frac{p+qx+ry}{1+x}$$

is increasing in both arguments. To this end assume for the sake of contradiction that the solution is not eventually in the interval $\left[0, \frac{q-p}{r}\right]$. Then one can see from (10.10) and the fact that now

$$q < \frac{q-p}{r},$$

that either

$$y_{2n} > \frac{q-p}{r} \quad \text{for} \quad n \geq 0 \tag{10.19}$$

or

$$y_{2n-1} > \frac{q-p}{r} \quad \text{for} \quad n \geq 0. \tag{10.20}$$

We will assume that (10.19) holds. The case where (10.20) holds is similar and will be omitted. Then from (10.10) we see that

$$y_{2n+2} - q = \frac{r}{1+y_{2n+1}}\left(y_{2n} - \frac{q-p}{r}\right) < r\left(y_{2n} - \frac{q-p}{r}\right)$$

and so

$$y_{2n+2} < ry_{2n} + p \quad \text{for} \quad n \geq 0.$$

Note that in this case $r < 1$ and so

$$\frac{q-p}{r} < y_{2n+2} < r^{n+1}y_0 + r^n p + \ldots + rp + p$$

which implies that

$$\frac{q-p}{r} \leq \frac{p}{1-r}.$$

This contradicts (10.14) and completes the proof of part *(b)* of Theorem 10.2.1.

10.2.2 Proof of Part *(c)* of Theorem 10.2.1

Here we assume that (10.7) holds and show that Eq(10.4) possesses unbounded solutions. To this end observe that

$$y_{n+1} = q + \frac{p + r(y_{n-1} - \frac{q}{r})}{1 + y_n} \quad \text{for} \quad n \geq 0$$

and so when (10.7) holds the interval $[q, \infty)$ is invariant. That is when

$$y_{-1}, y_0 \in [q, \infty),$$

then

$$y_n \geq q \quad \text{for} \quad n \geq 0.$$

Now the change of variables

$$y_n = q + z_n$$

transforms Eq(10.4) to Eq(10.9) and the conclusion of part *(c)* follows from Section 6.5 for appropriate initial conditions y_{-1} and y_0. More specifically the solutions of Eq(10.4) with initial conditions such that

$$q \leq y_{-1} < r - 1 \quad \text{and} \quad y_0 > r_{-1} + \frac{p + q(r-1)}{r - 1 - q}$$

have the property that the subsequences of even terms converge to ∞ while the subsequences of odd terms converge to q.

It should be mentioned that when (10.7) holds, the equilibrium of Eq(10.4) is a saddle point and so by the stable manifold theorem, in addition to unbounded solutions, Eq(10.4) possesses bounded solutions which in fact converge to the equilibrium of Eq(10.4).

10.3 The Case $B = 0: \; x_{n+1} = \frac{\alpha + \beta x_n + \gamma x_{n-1}}{A + C x_{n-1}}$

This is the (3, 2)-type Eq(10.2) which by the change of variables

$$x_n = \frac{A}{C} y_n$$

reduces to the difference equation

$$y_{n+1} = \frac{p + q y_n + r y_{n-1}}{1 + y_{n-1}}, \quad n = 0, 1, \ldots \tag{10.21}$$

where

$$p = \frac{\alpha C}{A^2}, \quad q = \frac{\beta}{A}, \quad \text{and} \quad r = \frac{\gamma}{A}.$$

This equation has not been investigated yet. Here we make some simple observations about linearized stability, invariant intervals, and convergence of solutions in certain regions of initial conditions.

The linearized equation of Eq(10.21) about its positive equilibrium

$$\overline{y} = \frac{q + r - 1 + \sqrt{(q + r - 1)^2 + 4p}}{2}$$

is

$$z_{n+1} - \frac{q}{1+\overline{y}} z_n + \frac{p - r + q\overline{y}}{(1+\overline{y})^2} z_n = 0, \quad n = 0, 1, \dots .$$

By applying Theorem 1.1.1 we see that $\overline{y}$ is locally asymptotically stable for all values of the parameters $p, q,$ and r.

10.3.1 Invariant Intervals

The following result, which can be established by direct calculation, gives a list of invariant intervals for Eq(10.21).

Recall that I is an invariant interval, if whenever,

$$y_N, y_{N+1} \in I \quad \text{for some integer} \quad N \geq 0,$$

then

$$y_n \in I \quad \text{for} \quad n \geq N.$$

Lemma 10.3.1 *Eq(10.21) possesses the following invariant intervals:*

(a)

$$\left[0, \frac{p}{1-q}\right] \quad \textit{when} \quad q < 1 \quad \textit{and} \quad p \geq r;$$

(b)

$$\left[0, \frac{r-p}{q}\right] \quad \textit{when} \quad q < 1 \quad \textit{and} \quad p \geq r(1-q).$$

10.3.2 Global Stability When $p \geq r$

Here we discuss the behavior of the solutions of Eq(10.21) when $p \geq r$.

Observe that

$$y_{n+1} = r + \frac{(p-r) + qy_n}{1 + y_{n-1}}, \quad n = 0, 1, \dots$$

and so when

$$p \geq r$$

$$y_n \geq r \quad \text{for} \quad n \geq 1.$$

Therefore the change of variables

$$y_n = r + z_n$$

transforms Eq(10.21) to the difference equation

$$z_{n+1} = \frac{p + qr - r + qz_n}{(1+r) + z_{n-1}}, \quad n = 1, 2, \ldots$$

which is of the form of the (2, 2)-type Eq(6.2) which was investigated in Section 6.3.

The following result is now a consequence of the above observations and the results in Section 6.3.

Theorem 10.3.1 *Assume*

$$p \geq r.$$

Then the following statements are true.

(a) Every positive solution of Eq(10.21) is bounded.

(b) The positive equilibrium of Eq(10.21) is globally asymptotically stable if one of the following conditions holds:

(i) $q < 1$;

(ii)

$$q \geq 1$$

and

$$\textit{either} \quad p + qr - r \leq q \quad \textit{or} \quad q < p + qr - r \leq 2(q+1).$$

10.3.3 Convergence of Solutions

Here we discuss the behavior of the solutions of Eq(10.21) when $p < r$.

Theorem 10.3.2 *(a) Assume that*

$$q < 1 \quad \textit{and} \quad r \geq \frac{p}{1-q}.$$

Then every solution of Eq(10.21) with initial conditions in the invariant interval

$$\left[0, \frac{r-p}{q}\right]$$

converges to the equilibrium $\overline{y}$.

(b) *Suppose that either*

$$0 \leq q-1 < r \quad \textit{and} \quad p < r$$

or

$$q < 1 \quad \textit{and} \quad p < r < \frac{p}{1-q}.$$

Set

$$K = \frac{pq + r^2 - pr}{q + r - p - q^2}.$$

Then every solution of Eq(10.21) with initial conditions in the invariant interval

$$\left[\frac{r-p}{q}, K\right]$$

converges to the equilibrium $\overline{y}$.

Proof.

(a) In view of Lemma 10.3.1 we see that when $y_{-1}, y_0 \in \left[0, \frac{r-p}{q}\right]$, then $y_n \in \left[0, \frac{r-p}{q}\right]$ for all $n \geq 0$. It is easy to check that $\overline{y} \in \left[0, \frac{r-p}{q}\right]$ and that in the interval $\left[0, \frac{r-p}{q}\right]$ the function $f(x, y)$ increases in both x and y. The result is now a consequence of Theorem 1.4.8.

(b) In view of Lemma 10.3.1 we see that when $y_{-1}, y_0 \in \left[\frac{r-p}{q}, K\right]$, then $y_n \in \left[\frac{r-p}{q}, K\right]$ for all $n \geq 0$. It is easy to check that $\overline{y} \in \left[\frac{r-p}{q}, K\right]$ and that in the interval $\left[\frac{r-p}{q}, K\right]$ the function $f(x, y)$ increases in x and decreases in y. The result is now a consequence of Theorem 1.4.5.

□

10.4 The Case $A = 0: \; x_{n+1} = \frac{\alpha+\beta x_n+\gamma x_{n-1}}{Bx_n+Cx_{n-1}}$

This is the $(3, 2)$-type Eq(10.3) which by the change of variables

$$x_n = \frac{\gamma}{C} y_n$$

reduces to the difference equation

$$y_{n+1} = \frac{r + py_n + y_{n-1}}{qy_n + y_{n-1}}, \quad n = 0, 1, \ldots \tag{10.22}$$

where

$$p = \frac{\beta}{\gamma}, \quad q = \frac{B}{C}, \quad \text{and} \quad r = \frac{\alpha C}{\gamma^2}.$$

This equation has not been investigated yet. Here we make some simple observations about linearized stability, invariant intervals, and the convergence of solutions in certain regions of initial conditions.

The linearized equation of Eq(10.22) about its positive equilibrium

$$\overline{y} = \frac{(1+p) + \sqrt{(1+p)^2 + 4r(1+q)}}{2(1+q)}$$

is

$$z_{n+1} - \frac{(p-q)\overline{y} - qr}{(q+1)(r+(p+1)\overline{y})} z_n + \frac{(p-q)\overline{y} + r}{(q+1)(r+(p+1)\overline{y})} z_{n-1} = 0, \quad n = 0, 1, \dots .$$

By applying Theorem 1.1.1 we see that $\overline{y}$ is locally asymptotically stable when either

$$q \leq pq + 1 + 3p \tag{10.23}$$

or

$$q > pq + 1 + 3p \quad \text{and} \quad F\left(\frac{2r}{q - pq - 1 - 3p}\right) > 0$$

where

$$F(u) = (q+1)u^2 - (p+1)u - r.$$

The second condition reduces to

$$q > pq + 1 + 3p. \tag{10.24}$$

It is interesting to note that the above condition for the local asymptotic stability of $\overline{y}$ requires the assumption that $r > 0$. When $r = 0$, as in Section 6.9, the inequality in (10.23) is strict and Condition (10.24) is void.

Concerning prime period-two solution, it follows from Section 2.5 that a period-two solution exists if and only if

$$p < 1, \quad q > 1 \quad \text{and} \quad 4r < (1-p)(q - pq - 3p - 1). \tag{10.25}$$

Furthermore in this case the prime period-two solution

$$\dots, \phi, \psi, \phi, \psi, \dots$$

is unique and the values ϕ and ψ are the positive roots of the quadratic equation

$$t^2 - (1-p)t + \frac{r + p(1-p)}{q-1} = 0.$$

10.4.1 Invariant Intervals

Here we present some results about invariant intervals for Eq(10.21). We consider the cases where $p = q$, $p > q$, and $p < q$.

Lemma 10.4.1 *Eq(10.21) possesses the following invariant intervals:*

(a)

$$[a, b] \quad \textit{when} \quad p = q \quad \textit{and}$$

a and b are positive numbers such that

$$r + (p+1)a \le (q+1)ab \le r + (p+1)b. \tag{10.26}$$

(b)

$$\left[\frac{p}{q}, \frac{qr}{p-q}\right] \quad \textit{when} \quad p > q \quad \textit{and} \quad \frac{p^2 - pq}{q^2} < r;$$

$$\left[\frac{qr}{p-q}, \frac{p}{q}\right] \quad \textit{when} \quad p > q \quad \textit{and} \quad \frac{p^2 - pq}{q^2} > r.$$

(c)

$$\left[1, \frac{r}{q-p}\right] \quad \textit{when} \quad p < q < p + r;$$

$$\left[\frac{r}{q-p}, 1\right] \quad \textit{when} \quad q > p + r.$$

Proof.

(a) It is easy to see that when $p = q$ the function

$$f(x, y) = \frac{r + px + y}{qx + y}$$

is decreasing in both arguments. Hence

$$a \le \frac{r + (p+1)b}{(q+1)b} = f(b, b) \le f(x, y) \le f(a, a) = \frac{r + (p+1)a}{(q+1)a} \le b.$$

(b) Clearly the function $f(x, y)$ is decreasing in both arguments when $y < \frac{qr}{p-q}$, and it is increasing in x and decreasing in y for $y \ge \frac{qr}{p-q}$.

First assume that $p > q$ and $\frac{p^2-pq}{q^2} < r$.

Then for $x, y \in \left[\frac{p}{q}, \frac{qr}{p-q}\right]$ we obtain

$$\frac{p}{q} = f\left(\frac{qr}{p-q}, \frac{qr}{p-q}\right) \leq f(x,y) \leq f\left(\frac{p}{q}, \frac{p}{q}\right) = \frac{qr + p(p+1)}{p(q+1)} \leq \frac{qr}{p-q}.$$

The inequalities

$$\frac{qr + p(p+1)}{p(q+1)} \leq \frac{qr}{p-q} \quad \text{and} \quad \frac{p}{q} < \frac{qr}{p-q}$$

are equivalent to the inequality $\frac{p^2 - pq}{q^2} < r$.

Next assume that $p > q$ and $\frac{p^2 - pq}{q^2} > r$.

For $x, y \in \left[\frac{qr}{p-q}, \frac{p}{q}\right]$, we obtain

$$\frac{(qr + p)(p - q) + pq^2 r}{q^3 r + p(p-q)} = f\left(\frac{qr}{p-q}, \frac{p}{q}\right) \leq f(x,y) \leq f\left(\frac{p}{q}, \frac{qr}{p-q}\right) = \frac{p}{q}.$$

The inequalities

$$\frac{(qr + p)(p - q) + pq^2 r}{q^3 r + p(p-q)} \geq \frac{qr}{p-q} \quad \text{and} \quad \frac{p}{q} > \frac{qr}{p-q}$$

follow from the inequality $\frac{p^2 - pq}{q^2} > r$.

(c) It is clear that the function $f(x, y)$ is decreasing in both arguments for $x < \frac{r}{q-p}$, and it is increasing in y and decreasing in x for $x \geq \frac{r}{q-p}$.

First, assume that $p < q < p + r$.

Using the decreasing character of f, we obtain

$$1 = f\left(\frac{r}{q-p}, \frac{r}{q-p}\right) \leq f(x,y) \leq f(1,1) = \frac{r + p + 1}{q + 1} \leq \frac{r}{q-p},$$

The inequalities

$$1 < \frac{r}{q-p} \quad \text{and} \quad \frac{r + p + 1}{q + 1} < \frac{r}{q-p}$$

are equivalent to the inequality $q < p + r$.

Next assume that $q > p + r$.

Using the decreasing character of f in x and the increasing character of f in y, we obtain

$$\frac{(p + r)(q - p) + r}{q(q-p) + r} = f\left(1, \frac{r}{q-p}\right) \leq f(x,y) \leq f\left(\frac{r}{p-q}, 1\right) = 1.$$

The inequalities

$$\frac{(p+r)(q-p)+r}{q(q-p)+r} \geq \frac{r}{q-p} \quad \text{and} \quad \frac{r}{q-p} < 1$$

follow from the inequality $q > p + r$.

□

10.4.2 Convergence of Solutions

Here we obtain some convergence results for Eq(10.22).

Theorem 10.4.1 *(a) Assume that $p = q$. Then every solution of Eq(10.22) with initial conditions in the invariant interval $[a, b]$, where $0 < a < b$ satisfy (10.26) converges to the equilibrium $\overline{y}$.*

(b) Assume that $p > q$ and $r > \frac{p^2-pq}{q^2}$. Then every solution of Eq(10.22) with initial conditions in the invariant interval $\left[\frac{p}{q}, \frac{qr}{p-q}\right]$ converges to the equilibrium $\overline{y}$.

(c) Assume that $p < q < p+r$. Then every solution of Eq(10.22) with initial conditions in the invariant interval $\left[1, \frac{r}{q-p}\right]$ converges to the equilibrium $\overline{y}$.

Proof. The proof is an immediate consequence of Lemma 10.4.1 and Theorem 1.4.7.

□

Theorem 10.4.2 *Assume that $p > q$, $r < \frac{p^2-pq}{q^2}$, and that either*

$$p \leq 1$$

or

$$p > 1 \quad \textit{and} \quad (q-1)[(p-1)^2(q-1) - 4q(1-p-qr)] \leq 0.$$

Then every solution of Eq(10.22) with initial conditions in the invariant interval $\left[\frac{qr}{p-q}, \frac{p}{q}\right]$ converges to the equilibrium $\overline{y}$.

Proof. The proof is an immediate consequence of Lemma 10.4.1 and Theorem 1.4.5.

□

Theorem 10.4.3 *Assume that $p < q$, $p+r < q$, and that Condition (10.25) is not satisfied. Then every solution of Eq(10.22) with initial conditions in the invariant interval $\left[\frac{r}{q-p}, 1\right]$ converges to the equilibrium $\overline{y}$.*

Proof. The proof is an immediate consequence of Lemma 10.4.1 and Theorem 1.4.6.

□

10.5 Open Problems and Conjectures

Open Problem 10.5.1 *We know that every solution $\{y_n\}_{n=-1}^{\infty}$ of Eq(10.4) converges to a not necessarily prime period-two solution*

$$\ldots, \phi, \psi, \phi, \psi, \ldots \tag{10.27}$$

if and only if

$$r = 1 + q.$$

(a) *Determine the set of initial conditions y_{-1} and y_0 for which $\phi \neq \psi$.*

(b) *Assume (10.27) is a prime period-two solution of Eq(10.4). Determine the set of initial conditions y_{-1} and y_0 for which $\{y_n\}_{n=-1}^{\infty}$ converges to (10.27).*

Open Problem 10.5.2 *Assume*

$$r > q + 1.$$

(a) *Find the set $\mathcal{B}$ of all initial conditions $(y_{-1}, y_0) \in (0, \infty) \times (0, \infty)$ such that the solutions $\{y_n\}_{n=-1}^{\infty}$ of Eq(10.4) are bounded.*

(b) *Let $(y_{-1}, y_0) \in \mathcal{B}$. Investigate the asymptotic behavior of $\{y_n\}_{n=-1}^{\infty}$.*

Conjecture 10.5.1 *Every positive solution of Eq(10.21) converges to the positive equilibrium of the equation.*

Open Problem 10.5.3 *For each of the equations listed below, determine the set $\mathcal{G}$ of all initial conditions $(x_{-1}, x_0) \in R \times R$ through which the equation is well defined for all $n \geq 0$, and for these initial points investigate the long-term behavior of the corresponding solution:*

$$x_{n+1} = \frac{1 + x_n + x_{n-1}}{1 + x_n} \tag{10.28}$$

$$x_{n+1} = \frac{1 + x_n + x_{n-1}}{1 + x_{n-1}} \tag{10.29}$$

$$x_{n+1} = \frac{1 + x_n + x_{n-1}}{x_n + x_{n-1}}. \tag{10.30}$$

Extend and generalize.

Conjecture 10.5.2 *Assume that*

$$p, q, r \in (0, \infty).$$

Then the following statements are true for Eq(10.22)

(a) Local stability of the positive equilibrium implies global stability.

(b) When Eq(10.22) has no prime period-two solutions the equilibrium $\overline{y}$ is globally asymptotically stable.

(c) When Eq(10.22) possesses a period-two solution, the equilibrium $\overline{y}$ is a saddle point.

(d) The period-two solution of Eq(10.22), when it exists, is locally asymptotically stable, but not globally.

Open Problem 10.5.4 *Assume that $\{p_n\}_{n=0}^{\infty}$, $\{q_n\}_{n=0}^{\infty}$, and $\{r_n\}_{n=0}^{\infty}$ are convergent sequences of nonnegative real numbers with finite limits,*

$$p = \lim_{n\to\infty} p_n, \quad q = \lim_{n\to\infty} q_n, \quad \text{and} \quad r = \lim_{n\to\infty} r_n.$$

Investigate the asymptotic behavior and the periodic nature of all positive solutions of each of the following three difference equations:

$$y_{n+1} = \frac{p_n + q_n y_n + r_n y_{n-1}}{1 + y_n}, \quad n = 0, 1, \ldots \tag{10.31}$$

$$y_{n+1} = \frac{p_n + q_n y_n + r_n y_{n-1}}{1 + y_{n-1}}, \quad n = 0, 1, \ldots \tag{10.32}$$

$$y_{n+1} = \frac{r_n + p_n y_n + y_{n-1}}{q_n y_n + y_{n-1}}, \quad n = 0, 1, \ldots . \tag{10.33}$$

Open Problem 10.5.5 *Assume that $\{p_n\}_{n=0}^{\infty}$, $\{q_n\}_{n=0}^{\infty}$, and $\{r_n\}_{n=0}^{\infty}$ are period-two sequences of nonnegative real numbers. Investigate the global character of all positive solutions of Eqs(10.31)-(10.33). Extend and generalize.*

Open Problem 10.5.6 *Assume that*

$$p, q, r \in [0, \infty) \quad \text{and} \quad k \in \{2, 3, \ldots\}.$$

Investigate the global behavior of all positive solutions of each of the following difference equations:

$$y_{n+1} = \frac{p + qy_n + ry_{n-k}}{1 + y_n}, \quad n = 0, 1, \ldots \tag{10.34}$$

$$y_{n+1} = \frac{p + qy_n + ry_{n-k}}{1 + y_{n-k}}, \quad n = 0, 1, \ldots \tag{10.35}$$

$$y_{n+1} = \frac{p + qy_n + ry_{n-1}}{qy_n + y_{n-k}}, \quad n = 0, 1, \ldots . \tag{10.36}$$

Chapter 11

The $(3,3)$-Type Equation $x_{n+1} = \frac{\alpha+\beta x_n+\gamma x_{n-1}}{A+Bx_n+Cx_{n-1}}$

In this chapter we discuss the $(3,3)$-type equation

$$x_{n+1} = \frac{\alpha + \beta x_n + \gamma x_{n-1}}{A + Bx_n + Cx_{n-1}}, \quad n = 0, 1, \ldots . \tag{11.1}$$

Please recall our classification convention in which all coefficients of this equation are now assumed to be positive and the initial conditions nonnegative.

This equation has not been investigated yet. Here we make some simple observations about linearized stability, invariant intervals, and the convergence of solutions in certain regions of initial conditions.

11.1 Linearized Stability Analysis

Eq(11.1) has a unique equilibrium which is the positive solution of the quadratic equation

$$(B + C)\overline{x}^2 - (\beta + \gamma - A)\overline{x} - \alpha = 0.$$

That is,

$$\overline{x} = \frac{\beta + \gamma - A + \sqrt{(A - \beta - \gamma)^2 + 4\alpha(B + C)}}{2(B + C)}.$$

The linearized equation about $\overline{x}$ is (see Section 2.3)

$$\begin{aligned} z_{n+1} &- \frac{(\beta A - B\alpha) + (\beta C - B\gamma)\overline{x}}{A^2 + \alpha(B + C) + (B + C)(A + \beta + \gamma)\overline{x}} z_n \\ &- \frac{(\gamma A - C\alpha) - (\beta C - \gamma B)\overline{x}}{A^2 + \alpha(B + C) + (B + C)(A + \beta + \gamma)\overline{x}} z_{n-1} = 0, \quad n = 0, 1, \ldots . \end{aligned} \tag{11.2}$$

The equilibrium $\overline{x}$ is locally asymptotically stable if Conditions (2.13), (2.14), and (2.15) are satisfied.

Eq(11.1) has a prime period-two solution

$$\ldots, \phi, \psi, \phi, \psi, \ldots$$

if, see Section 2.5, Conditions (2.31) and (2.32) are satisfied; that is,

$$\gamma > \beta + A, \quad B > C, \quad \text{and} \quad \alpha < \frac{(\gamma - \beta - A)[B(\gamma - \beta - A) - C(\gamma + 3\beta - A)]}{4C^2}. \tag{11.3}$$

Furthermore, when (11.3), holds the values of ϕ and ψ are the positive roots of the quadratic equation

$$t^2 + \frac{\gamma - \beta - A}{C} t + \frac{\alpha C + \beta(\gamma - \beta - A)}{C(B - C)} = 0. \tag{11.4}$$

The local asymptotic stability of the two cycle

$$\ldots, \phi, \psi, \phi, \psi, \ldots$$

is discussed in Section 2.6.

11.2 Invariant Intervals

Here we obtain several invariant intervals for Eq(11.1). These intervals are derived by analyzing the intervals where the function

$$f(x, y) = \frac{\alpha + \beta x + \gamma y}{A + Bx + Cy}$$

is increasing or decreasing in x and y.

If we set

$$L = \beta A - \alpha B, \quad M = \beta C - \gamma B, \quad \text{and} \quad N = \gamma A - \alpha C$$

then clearly

$$f_x = \frac{L + My}{(A + Bx + Cy)^2} \quad \text{and} \quad f_y = \frac{N - My}{(A + Bx + Cy)^2}.$$

Recall that an interval I is an invariant interval for Eq(11.1) if whenever,

$$y_N, y_{N+1} \in I \quad \text{for some integer} \quad N \geq 0,$$

then

$$y_n \in I \quad \text{for} \quad n \geq N.$$

Once we have an invariant interval for Eq(11.1) then we may be able to obtain a convergence result for the solutions of Eq(11.1) by testing whether the hypotheses of one of the four Theorems 1.4.5-1.4.8 are satisfied.

Lemma 11.2.1 *(i) Assume that*

$$\frac{\gamma}{C} = \frac{\beta}{B} > \frac{\alpha}{A}.$$

Then $\left[\frac{\alpha}{A}, \frac{\beta}{B}\right]$ *is an invariant interval for Eq(11.1) and in this interval the function* $f(x,y)$ *is increasing in both variables.*

(ii) Assume that

$$\frac{\beta}{B} > \frac{\gamma}{C} > \frac{\alpha}{A} \quad \text{and} \quad \frac{\gamma}{C} \leq \frac{N}{M}.$$

Then $\left[\frac{\alpha}{A}, \frac{\gamma}{C}\right]$ *is an invariant interval for Eq(11.1) and in this interval the function* $f(x,y)$ *is increasing in both variables.*

(iii) Assume that

$$\frac{\gamma}{C} > \frac{\beta}{B} > \frac{\alpha}{A} \quad \text{and} \quad \frac{\beta+\gamma}{B+C} \leq \frac{\alpha B - \beta A}{\beta C - \gamma B}.$$

Then $\left[\frac{\alpha}{A}, \frac{\beta+\gamma}{B+C}\right]$ *is an invariant interval for Eq(11.1) and in this interval the function* $f(x,y)$ *is increasing in both variables.*

(iv) Assume that

$$\frac{\gamma}{C} = \frac{\beta}{B} < \frac{\alpha}{A}.$$

Then $\left[\frac{\beta}{B}, \frac{\alpha}{A}\right]$ *is an invariant interval for Eq(11.1) and in this interval the function* $f(x,y)$ *is decreasing in both variables.*

(v) Assume that

$$\frac{\gamma A - \alpha C}{\beta C - \gamma B} \geq \frac{\alpha}{A} > \frac{\beta}{B} > \frac{\gamma}{C}.$$

Then $\left[0, \frac{\gamma A - \alpha C}{\beta C - \gamma B}\right]$ *is an invariant interval for Eq(11.1) and in this interval the function* $f(x,y)$ *is decreasing in both variables.*

(vi) Assume that

$$\frac{\beta}{B} < \frac{\gamma}{C} < \frac{\alpha}{A} < \frac{\gamma A - \alpha C}{\beta C - \gamma B}.$$

Then $\left[0, \frac{\alpha}{A}\right]$ *is an invariant interval for Eq(11.1) and in this interval the function* $f(x,y)$ *is decreasing in both variables.*

(vii) Assume that

$$\frac{\alpha}{A} = \frac{\beta}{B} > \frac{\gamma}{C}.$$

Then $\left[0, \frac{\alpha}{A}\right]$ *is an invariant interval for Eq(11.1) and in this interval the function* $f(x,y)$ *is increasing in* x *and decreasing in* y*.*

(viii) Assume that

$$\frac{\beta}{B} > \frac{\alpha}{A} \geq \frac{\gamma}{C}.$$

Then $\left[0, \frac{\beta}{B}\right]$ is an invariant interval for Eq(11.1) and in this interval the function $f(x,y)$ is increasing in x and decreasing in y.

(ix) Assume that

$$\frac{\alpha}{A} = \frac{\beta}{B} < \frac{\gamma}{C}.$$

Then $\left[0, \frac{\gamma}{C}\right]$ is an invariant interval for Eq(11.1) and in this interval the function $f(x,y)$ is increasing in y and decreasing in x.

(x) Assume that

$$\frac{\beta}{B} < \frac{\alpha}{A} \leq \frac{\gamma}{C}.$$

Then $\left[0, \frac{\gamma}{C}\right]$ is an invariant interval for Eq(11.1) and in this interval the function $f(x,y)$ is increasing in y and decreasing in x.

Proof.

(i) The condition $\frac{\gamma}{C} = \frac{\beta}{B} > \frac{\alpha}{A}$ is equivalent to $M = 0$ and $L > 0$, which in turn imply that $N > 0$. Thus, in this case the function f is increasing in both variables for all values of x and y and

$$\frac{\alpha}{A} = f(0,0) \leq f(x,y) \leq f(\frac{\beta}{B}, \frac{\beta}{B}) \leq \frac{\beta+\gamma}{B+C} = \frac{\beta}{B}.$$

Here we used the fact that under the condition

$$\frac{\gamma}{C} = \frac{\beta}{B} > \frac{\alpha}{A}$$

the function $f(u,u)$ is increasing and consequently

$$f(u,u) \leq \frac{\beta+\gamma}{B+C} = \frac{\beta}{B}.$$

(ii) The condition $\frac{\beta}{B} > \frac{\gamma}{C} > \frac{\alpha}{A}$ is equivalent to $M > 0, L > 0$, and $N > 0$. Thus, in this case the function f is increasing in x, for every x. In addition, this function is increasing in y for $x < \frac{N}{M}$. Thus, assuming $\frac{\gamma}{C} \leq \frac{N}{M}$ we obtain

$$\frac{\alpha}{A} = f(0,0) \leq f(x,y) \leq f(\frac{N}{M}, \frac{N}{M}) = \frac{\gamma}{C},$$

for all $x, y \in \left[\frac{\alpha}{A}, \frac{\gamma}{C}\right]$.

(iii) The condition $\frac{\gamma}{C} > \frac{\beta}{B} > \frac{\alpha}{A}$ is equivalent to $M < 0$ and $L > 0$, which in turn imply that $N > 0$. Thus, in this case the function f is increasing in y, for every y. In addition, this function is increasing in x for $y < -\frac{L}{M}$. Thus assuming $\frac{\beta+\gamma}{B+C} \leq -\frac{L}{M}$, we obtain,

$$\frac{\alpha}{A} = f(0,0) \leq f(x,y) \leq f\left(\frac{\beta+\gamma}{B+C}, \frac{\beta+\gamma}{B+C}\right) \leq \frac{\beta+\gamma}{B+C},$$

for all $x, y \in \left[\frac{\alpha}{A}, \frac{\beta+\gamma}{B+C}\right]$.

(iv) The condition $\frac{\gamma}{C} = \frac{\beta}{B} < \frac{\alpha}{A}$ is equivalent to $M = 0$ and $L < 0$, which in turn imply that $N < 0$. Thus, in this case the function f is decreasing in both variables for all values of x and y and

$$\frac{\beta}{B} \leq f\left(\frac{\alpha}{A}, \frac{\alpha}{A}\right) \leq f(x,y) \leq f(0,0) = \frac{\alpha}{A}.$$

Here we used the fact that under the condition

$$\frac{\gamma}{C} = \frac{\beta}{B} < \frac{\alpha}{A}$$

the function $f(u,u)$ is decreasing and consequently

$$f(u,u) \geq \frac{\beta+\gamma}{B+C} = \frac{\beta}{B}.$$

(v) The condition $\frac{\alpha}{A} > \frac{\beta}{B} > \frac{\gamma}{C}$ is equivalent to $M > 0$ and $L < 0$, which imply that $N < 0$. Thus, in this case the function f is decreasing in y, for every y. In addition, this function is decreasing in x for $y < -\frac{L}{M}$. Thus assuming $\frac{\alpha}{A} \leq -\frac{L}{M}$, we obtain

$$0 \leq f(x,y) \leq f(0,0) = \frac{\alpha}{A} \leq -\frac{L}{M},$$

for all $x, y \in \left[0, -\frac{L}{M}\right]$.

(vi) The condition $\frac{\alpha}{A} > \frac{\gamma}{C} > \frac{\beta}{B}$ is equivalent to $M < 0, L < 0$, and $N < 0$. Thus, in this case the function f is decreasing in x, for every x. In addition, this function is decreasing in y for $x < \frac{N}{M}$. Thus assuming $\frac{\alpha}{A} \leq \frac{N}{M}$, we obtain

$$0 \leq f(x,y) \leq f(0,0) = \frac{\alpha}{A},$$

for all $x, y \in \left[0, \frac{\alpha}{A}\right]$.

(vii) The condition $\frac{\alpha}{A} = \frac{\beta}{B} > \frac{\gamma}{C}$ is equivalent to $M > 0$ and $L = 0$, which in turn imply that $N < 0$. Thus, in this case the function f is increasing in x and decreasing in y for all values of x and y. Using this we obtain,

$$0 \leq f(x,y) \leq f\left(\frac{\alpha}{A}, 0\right) = \frac{\alpha}{A},$$

for all $x, y \in \left[0, \frac{\alpha}{A}\right]$.

(viii) The condition $\frac{\beta}{B} > \frac{\alpha}{A} > \frac{\gamma}{C}$ is equivalent to $M > 0, L > 0$, and $N < 0$. Thus, in this case the function f is increasing in x and decreasing in y for all values of x and y. Using this we obtain,

$$0 \leq f(x,y) < f\left(\frac{\beta}{B}, 0\right) = \frac{\beta}{B},$$

for all $x, y \in \left[0, \frac{\beta}{B}\right]$.

(ix) The condition $\frac{\alpha}{A} = \frac{\beta}{B} < \frac{\gamma}{C}$ is equivalent to $M < 0$ and $L = 0$, which in turn imply that $N > 0$. Thus, in this case the function f is decreasing in x and increasing in y for all values of x and y. Using this we obtain,

$$0 \leq f(x,y) \leq f\left(0, \frac{\gamma}{C}\right) = \frac{\gamma}{C},$$

for all $x, y \in \left[0, \frac{\gamma}{C}\right]$.

(x) The condition $\frac{\beta}{B} < \frac{\alpha}{A} \leq \frac{\gamma}{C}$ is equivalent to $M < 0, L < 0$, and $N \geq 0$. Thus, in this case the function f is increasing in x and decreasing in y for all values of x and y. Using this we obtain,

$$0 \leq f(x,y) < f\left(0, \frac{\gamma}{C}\right) \leq \frac{\gamma}{C},$$

for all $x, y \in \left[0, \frac{\gamma}{C}\right]$. Here we use the fact that the function $f(0,v)$ is increasing in v when $N > 0$ and so

$$f(0,v) \leq \frac{\gamma}{C} = \lim_{v\to\infty} f(0,v),$$

for all v.

□

The next table summarizes more complicated invariant intervals with the signs of the partial derivatives f_x and f_y.

Case	Invariant intervals	Signs of derivatives
1	$[0,\frac{N}{M}]$ if $\frac{\alpha M+(\beta+\gamma)N}{AM+(B+C)N}<\frac{N}{M}$ $[\frac{N}{M},K_1]$ if $\frac{\alpha M+(\beta+\gamma)N}{AM+(B+C)N}>\frac{N}{M}$	$f_x>0$ $f_y>0$ if $x<\frac{N}{M}$; $f_y<0$ if $x>\frac{N}{M}$
2	$[0,\frac{\beta-A+\sqrt{(\beta-A)^2+4B\alpha}}{2B}]$	$f_x>0$ $f_y<0$
3	$[0,\frac{\beta+\gamma-A+\sqrt{(\beta+\gamma-A)^2+4\alpha(B+C)}}{2(B+C)}]$	$f_x>0$ $f_y>0$
4	$[0,-\frac{L}{M}]$ if $\frac{\alpha M-(\beta+\gamma)L}{AM-(B+C)L}<-\frac{L}{M}$ $[-\frac{L}{M},K_2]$ if $\frac{\beta MK+\alpha M-\gamma L}{BMK+AM-CL}>-\frac{L}{M}$	$f_x>0$ if $y<-\frac{L}{M}$; $f_x<0$ if $y>-\frac{L}{M}$ $f_y>0$
5	$[0,\frac{\beta}{B}]$	$f_x>0$ $f_y<0$
6	$[0,\frac{\gamma}{C}]$	$f_x<0$ $f_y>0$
7	$[\frac{\alpha M-L(\beta+\gamma)}{AM-L(B+C)},-\frac{L}{M}]$ if $\frac{\alpha M-L(\beta+\gamma)}{AM-L(B+C)}<-\frac{L}{M}$ $[-\frac{L}{M},K_3]$ if $\frac{\alpha M-\beta L+\gamma MK}{AM-BL+CMK}>-\frac{L}{M}$	$f_x>0$ if $y>-\frac{L}{M}$; $f_x<0$ if $y<-\frac{L}{M}$ $f_y<0$
8	$[0,\frac{\alpha}{A}]$	$f_x<0$ $f_y<0$
9	$[\frac{\alpha M+(\beta+\gamma)N}{AM+(B+C)N},\frac{N}{M}]$ if $\frac{\alpha M+(\beta+\gamma)N}{AM+(B+C)N}<\frac{N}{M}$ $[\frac{N}{M},K_4]$ if $\frac{\alpha M+\beta K+\gamma NM}{AM+BK+CNM}>\frac{N}{M}$	$f_x<0$ $f_y>0$ if $x>\frac{N}{M}$; $f_y<0$ if $x<\frac{N}{M}$

where

$$K_1=\frac{(\beta-A)M-CN+\sqrt{((\beta-A)M-NC)^2+4BM(\alpha M+\gamma N)}}{2BM},$$

$$K_2=\frac{(A-\gamma)M-BL+\sqrt{((A-\gamma)M-BL)^2+4CM(\alpha M-\beta L)}}{-2CM},$$

$$K_3=\frac{(\beta-A)M+LC+\sqrt{((\beta-A)M+LC)^2+4BM(\alpha M-\gamma L)}}{2BM},$$

$$K_4=\frac{(\gamma-A)M-BN+\sqrt{((\gamma-A)M-BN)^2+4BM(\alpha M+\beta N)}}{2CM}.$$

11.3 Convergence Results

The following result is a consequence of Lemma 11.2.1 and Theorems 1.4.5-1.4.8.

Theorem 11.3.1 *Let $\{x_n\}$ be a solution of Eq(11.1). Then in each of the following cases*

$$\lim_{n\to\infty}x_n=\overline{x}.$$

(i)

$$\frac{\gamma}{C} = \frac{\beta}{B} > \frac{\alpha}{A} \quad \text{and} \quad x_{-1}, x_0 \in \left[\frac{\alpha}{A}, \frac{\beta}{B}\right];$$

(ii)

$$\frac{\beta}{B} > \frac{\gamma}{C} > \frac{\alpha}{A}, \quad \frac{\gamma}{C} \leq \frac{\gamma A - \alpha C}{\beta C - \gamma B}, \quad \text{and} \quad x_{-1}, x_0 \in \left[\frac{\alpha}{A}, \frac{\gamma}{C}\right];$$

(iii)

$$\frac{\gamma}{C} > \frac{\beta}{B} > \frac{\alpha}{A}, \quad \frac{\beta+\gamma}{B+C} \leq \frac{\alpha B - \beta A}{\beta C - \gamma B}, \quad \text{and} \quad x_{-1}, x_0 \in \left[\frac{\alpha}{A}, \frac{\beta+\gamma}{B+C}\right];$$

(iv)

$$\frac{\gamma}{C} = \frac{\beta}{B} < \frac{\alpha}{A} \quad \text{and} \quad x_{-1}, x_0 \in \left[\frac{\beta}{B}, \frac{\alpha}{A}\right];$$

(v)

$$\frac{\gamma A - \alpha C}{\beta C - \gamma B} \geq \frac{\alpha}{A} > \frac{\beta}{B} > \frac{\gamma}{C} \quad \text{and} \quad x_{-1}, x_0 \in \left[0, \frac{\gamma A - \alpha C}{\beta C - \gamma B}\right];$$

(vi)

$$\frac{\beta}{B} < \frac{\gamma}{C} < \frac{\alpha}{A} < \frac{\gamma A - \alpha C}{\beta C - \gamma B} \quad \text{and} \quad x_{-1}, x_0 \in \left[0, \frac{\alpha}{A}\right];$$

(vii) At least one of the following three conditions is not satisfied and $x_{-1}, x_0 \in \left[0, \frac{\alpha}{A}\right]$*:*

$$\beta > \gamma + A,$$

$$B < C,$$

and

$$(C-B)[(C-B)(\beta-\gamma-A)^2 - 4B(\alpha B + \gamma(\beta-\gamma-A))] > 0,$$

or

$$\frac{\alpha}{A} = \frac{\beta}{B} > \frac{\gamma}{C};$$

(viii) At least one of the following three conditions is not satisfied and $x_{-1}, x_0 \in \left[0, \frac{\beta}{B}\right]$*:*

$$\beta > \gamma + A,$$

$$B < C,$$

and

$$(C-B)[(C-B)(\beta-\gamma-A)^2 - 4B(\alpha B + \gamma(\beta-\gamma-A))] > 0,$$

or

$$\frac{\beta}{B} > \frac{\alpha}{A} \geq \frac{\gamma}{C};$$

(ix) Condition (11.3) is not satisfied,

$$\frac{\alpha}{A} = \frac{\beta}{B} < \frac{\gamma}{C} \quad \textit{and} \quad x_{-1}, x_0 \in \left[0, \frac{\gamma}{C}\right];$$

(x) Condition (11.3) is not satisfied,

$$\frac{\beta}{B} < \frac{\alpha}{A} \leq \frac{\gamma}{C} \quad \textit{and} \quad x_{-1}, x_0 \in \left[0, \frac{\gamma}{C}\right].$$

Proof. The proofs of statements *(i-iii)* follow by applying Lemma 11.2.1 *(i-iii)* and Theorem 1.4.8.

The proofs of statements *(iv-vi)* follow by applying Lemma 11.2.1 *(iv-vi)* and Theorem 1.4.7.

The proofs of statements *(vii)* and *(viii)* follow by applying Lemma 11.2.1 *(vii)* and *(viii)* respectively, Theorem 1.4.5, and by observing that the only solution of the system of equations

$$m = f(m, M), \quad M = f(M, m);$$

that is,

$$m = \frac{\alpha + \beta m + \gamma M}{A + Bm + CM}, \quad M = \frac{\alpha + \beta M + \gamma m}{A + BM + Cm}$$

is $m = M$.

The proofs of statements *(ix)* and *(x)* follow by applying Lemma 11.2.1 *(ix)* and *(x)* respectively, Theorem 1.4.6, and by observing that Eq(11.1) does not possess a prime period-two solution.

□

11.3.1 Global Stability

Theorem 1.4.2 and linearized stability imply the following global stability result:

Theorem 11.3.2 *Assume that*

$$\frac{\gamma}{C} < \min\left\{\frac{\beta}{B}, \frac{\alpha}{A}\right\}.$$

Then the equilibrium $\overline{x}$ of Eq(11.1) is globally asymptotically stable.

By applying Theorem 1.4.3 to Eq(11.1) we obtain the following global asymptotic result.

Theorem 11.3.3 *Assume that*

$$B \le C,\ \gamma B > \beta C,\ \alpha B > \beta A, \quad \text{and} \quad \gamma A > \alpha C.$$

Then the equilibrium $\overline{x}$ of Eq(11.1) is globally asymptotically stable.

The following global stability result is a consequence of linearized stability and the Trichotomy Theorem 1.4.4.

Theorem 11.3.4 *(a) Assume that*

$$\alpha C \ge \gamma A,\ \beta A \ge \alpha B, \quad \text{and} \quad 3\gamma B \ge \beta C \ge \gamma B.$$

Then the equilibrium $\overline{x}$ of Eq(11.1) is globally asymptotically stable.

(b) Assume that

$$\alpha B \ge \beta A,\ \gamma A \ge \alpha C, \quad \text{and} \quad 3\beta C \ge \gamma B \ge \beta C.$$

Then the equilibrium $\overline{x}$ of Eq(11.1) is globally asymptotically stable.

11.4 Open Problems and Conjectures

Conjecture 11.4.1 *Assume*

$$\alpha, \beta, \gamma, A, B, C \in (0, \infty).$$

Show that every positive solution of Eq(11.1) is bounded.

Conjecture 11.4.2 *Assume*

$$\alpha, \beta, \gamma, A, B, C \in (0, \infty)$$

and that Eq(11.1) has no positive prime period-two solution. Show that every positive solution of Eq(11.1) converges to the positive equilibrium.

Conjecture 11.4.3 *Assume*

$$\alpha, \beta, \gamma, A, B, C \in (0, \infty)$$

and that Eq(11.1) has a positive prime period-two solution. Show that the positive equilibrium of Eq(11.1) is a saddle point and the period-two solution is locally asymptotically stable.

Conjecture 11.4.4 *Assume*

$$\alpha, \beta, \gamma, A, B, C \in (0, \infty).$$

Show that Eq(11.1) cannot have a prime period-k solution for any $k \geq 3$.

Open Problem 11.4.1 *Assume* $\{\alpha_n\}_{n=0}^{\infty}$, $\{\beta_n\}_{n=0}^{\infty}$, $\{\gamma_n\}_{n=0}^{\infty}$, $\{A_n\}_{n=0}^{\infty}$, $\{B_n\}_{n=0}^{\infty}$, $\{C_n\}_{n=0}^{\infty}$ *are convergent sequences of nonnegative real numbers with finite limits. Investigate the asymptotic behavior and the periodic nature of all positive solutions of the difference equation*

$$x_{n+1} = \frac{\alpha_n + \beta_n x_n + \gamma_n x_{n-1}}{A_n + B_n x_n + C_n x_{n-1}}, \quad n = 0, 1, \ldots . \tag{11.5}$$

Open Problem 11.4.2 *Assume that the parameters in Eq(11.5) are period-two sequences of positive real numbers. Investigate the asymptotic behavior and the periodic nature of all positive solutions of Eq(11.5).*

Open Problem 11.4.3 *Assume that*

$$a, A, a_i, A_i \in [0, \infty) \quad \textit{for} \quad i = 0, 1, \ldots, k. \tag{11.6}$$

Obtain necessary and sufficient conditions so that every positive solution of the equation

$$x_{n+1} = \frac{a + a_0 x_n + \ldots + a_k x_{n-k}}{A + A_0 x_n + \ldots + A_k x_{n-k}}, \quad n = 0, 1, \ldots . \tag{11.7}$$

has a finite limit.

Open Problem 11.4.4 *Assume that (11.6) holds. Obtain necessary and sufficient conditions so that every positive solution of Eq(11.7) converges to a period-two solution.*

Open Problem 11.4.5 *Assume that (11.6) holds. Obtain necessary and sufficient conditions so that every positive solution of Eq(11.7) is periodic with period-k,* $k \geq 2$. *In particular, investigate the cases where* $2 \leq k \leq 8$.

Open Problem 11.4.6 *Let*

$$\alpha, \beta, \gamma, \delta \in [0, \infty) \quad \textit{with} \quad \gamma + \delta > 0.$$

Investigate the boundedness character, the periodic nature, and the asymptotic behavior of all positive solutions of the difference equation

$$x_{n+1} = \frac{\alpha x_n + \beta x_{n-1}x_{n-2}}{\gamma x_n x_{n-1} + \delta x_{n-2}}, \quad n = 0, 1, \ldots .$$

(See [8] and [61].)

Conjecture 11.4.5 *Let $k \geq 2$ be a positive integer. Show that every positive solution of the difference equation*

$$x_{n+1} = \frac{x_n + \ldots + x_{n-k} + x_{n-k-1}x_{n-k-2}}{x_n x_{n-1} + x_{n-2} + \ldots + x_{n-k-2}}, n = 0, 1, \ldots$$

converges to 1. *Extend and generalize.*

(See [8], [47], and [60].)

Conjecture 11.4.6 *Let*

$$a_{ij} \in (0, \infty) \quad \textit{for} \quad i, j \in \{1, 2, 3\}.$$

Show that every positive solution of the system

$$\left.\begin{aligned} x_{n+1} &= \tfrac{a_{11}}{x_n} + \tfrac{a_{12}}{y_n} + \tfrac{a_{13}}{z_n} \\ y_{n+1} &= \tfrac{a_{21}}{x_n} + \tfrac{a_{22}}{y_n} + \tfrac{a_{23}}{z_n} \\ z_{n+1} &= \tfrac{a_{31}}{x_n} + \tfrac{a_{32}}{y_n} + \tfrac{a_{33}}{z_n} \end{aligned}\right\}, n = 0, 1, \ldots \tag{11.8}$$

converges to a period-two solution. Extend to equations with real parameters.

(See [32] for the two dimensional case.)

Conjecture 11.4.7 *Assume that*

$$p, q \in [0, \infty).$$

(a) Show that every positive solution of the equation

$$x_{n+1} = \frac{px_{n-1} + x_{n-2}}{q + x_{n-2}}, \quad n = 0, 1, \ldots$$

converges to a period-two solution if and only if

$$p = 1 + q.$$

(b) Show that when

$$p < 1 + q$$

every positive solution has a finite limit.

(c) Show that when

$$p > 1 + q$$

the equation possesses positive unbounded solutions.

(See [62].)

Conjecture 11.4.8 *Assume that*

$$p \in (0, \infty).$$

(a) Show that every positive solution of the equation

$$x_{n+1} = \frac{x_{n-1}}{x_{n-1} + px_{n-2}}, \quad n = 0, 1, \ldots$$

converges to a period-two solution if

$$p \geq 1.$$

(b) Show that when

$$p < 1$$

the positive equilibrium $\overline{x} = \frac{1}{1+p}$ is globally asymptotically stable.

Conjecture 11.4.9 *Assume that*

$$p, q \in [0, \infty).$$

(a) *Show that every positive solution of the equation*

$$x_{n+1} = \frac{x_{n-1} + p}{x_{n-2} + q}, \quad n = 0, 1, \ldots$$

converges to a period-two solution if and only if

$$q = 1.$$

(b) *Show that when*

$$q > 1$$

the positive equilibrium of the equation is globally asymptotically stable.

(c) *Show that when*

$$q < 1$$

the equation possesses positive unbounded solutions.

(See [62].)

Conjecture 11.4.10 *Show that every positive solution of the equation*

$$x_{n+1} = \frac{x_n + x_{n-2}}{x_{n-1}}, \quad n = 0, 1, \ldots$$

converges to a period-four solution.

(See [62].)

Conjecture 11.4.11 *Show that the difference equation*

$$x_{n+1} = \frac{p + x_{n-2}}{x_n}, \quad n = 0, 1, \ldots$$

has the following trichotomy character:

(i) *When* $p > 1$ *every positive solution converges to the positive equilibrium.*

(ii) *When* $p = 1$ *every positive solution converges to a period-five solution.*

(iii) *When* $p < 1$ *there exist positive unbounded solutions.*

(See [62].)

Conjecture 11.4.12 *Show that every positive solution of the equation*

$$x_{n+1} = \frac{1 + x_n}{x_{n-1} + x_{n-2}}, \quad n = 0, 1, \ldots$$

converges to a period-six solution of the form

$$\ldots, \phi, \psi, \frac{\psi}{\phi}, \frac{1}{\phi}, \frac{1}{\psi}, \frac{\phi}{\psi}, \ldots$$

where

$$\phi, \psi \in (0, \infty).$$

(See [62].)

Open Problem 11.4.7 *Consider the difference equation*

$$x_{n+1} = \frac{\alpha + \beta x_n + \gamma x_{n-1} + \delta x_{n-2}}{A + B x_n + D x_{n-2}}, n = 0, 1, \ldots \tag{11.9}$$

with nonnegative parameters and nonnegative initial conditions such that $B + D > 0$ *and*

$$A + B x_n + D x_{n-2} > 0 \quad \textit{for} \quad n \geq 0.$$

Then one can show that Eq(11.9) has prime period-two solutions if and only if

$$\gamma = \beta + \delta + A. \tag{11.10}$$

(a) *In addition to (11.10), what other conditions are needed so that every positive solution of Eq(11.9) converges to a period-two solution?*

(b) *Assume*

$$\gamma < \beta + \delta + A.$$

What other conditions are needed so that every positive solution of Eq(11.9) converges to the positive equilibrium?

(c) *Assume*

$$\gamma > \beta + \delta + A.$$

What other conditions are needed so that Eq(11.9) possesses unbounded solutions?

(See [6].)

Conjecture 11.4.13 *Show that every positive solution of each of the two equations*

$$x_{n+1} = \frac{1 + x_{n-1} + x_{n-2}}{x_n}, \quad n = 0, 1, \ldots$$

and

$$x_{n+1} = \frac{1 + x_{n-1} + x_{n-2}}{x_{n-2}}, \quad n = 0, 1, \ldots$$

converges to a period-two solution.

(See [6].)

Conjecture 11.4.14 *Show that every positive solution of the equation*

$$x_{n+1} = \frac{1 + x_{n-1}}{1 + x_n + x_{n-2}}, \quad n = 0, 1, \ldots$$

converges to a period-two solution.

(See [6].)

Conjecture 11.4.15 *Show that every positive solution of the equation*

$$x_{n+1} = \frac{x_{n-2}}{x_{n-1} + x_{n-2} + x_{n-3}}, \quad n = 0, 1, \ldots$$

converges to a period-three solution.

Conjecture 11.4.16 *Consider the difference equation*

$$x_{n+1} = \frac{p + qx_n + rx_{n-2}}{x_{n-1}}, \quad n = 0, 1, \ldots$$

where

$$p \in [0, \infty) \quad \text{and} \quad q, r \in (0, \infty).$$

(a) Show that when

$$q = r$$

every positive solution converges to a period-four solution.

(b) Show that when

$$q > r$$

every positive solution converges to the positive equilibrium.

(c) Show that when

$$q < r$$

the equation possesses positive unbounded solutions.

(See [62].)

Open Problem 11.4.8 *Determine whether every positive solution of each of the following equations converges to a periodic solution of the corresponding equation:*

(a)

$$x_{n+1} = \frac{x_{n-2} + x_{n-3}}{x_n}, \quad n = 0, 1, \ldots$$

(b)

$$x_{n+1} = \frac{1 + x_{n-3}}{1 + x_n + x_{n-3}}, \quad n = 0, 1, \ldots$$

(c)

$$x_{n+1} = \frac{1 + x_n + x_{n-3}}{x_{n-2}}, \quad n = 0, 1, \ldots$$

(d)

$$x_{n+1} = \frac{x_{n-1}}{x_n + x_{n-2} + x_{n-3}}, \quad n = 0, 1, \ldots .$$

Open Problem 11.4.9 *Assume that k is a positive integer and that*

$$A, B_0, B_1, \ldots B_k \in [0, \infty) \quad \textit{with} \quad \sum_{i=0}^{k} B_i > 0.$$

Obtain necessary and sufficient conditions on the parameters A and $B_0, B_1, \ldots B_k$ so that every positive solution of the difference equation

$$x_{n+1} = \frac{x_{n-k}}{A + B_0 x_n + \ldots + B_k x_{n-k}}, \quad n = 0, 1, \ldots$$

converges to a period-$(k+1)$ solution of this equation.

(See [6].)

Open Problem 11.4.10 *Assume that k is a positive integer. Then we can easily see that every nonnegative solution of the equation*

$$x_{n+1} = \frac{x_{n-k}}{1 + x_n + \ldots + x_{n-k+1}}, \quad n = 0, 1, \ldots \tag{11.11}$$

converges to a period-$(k+1)$ solution of the form

$$\ldots, \underbrace{0, 0, ..., 0}_{k-terms}, \phi, \ldots \tag{11.12}$$

with $\phi \geq 0$.

(a) Does Eq(11.11) possess a positive solution which converges to zero?

(b) Determine, in terms of the nonnegative initial conditions $x_{-k}, \ldots, x_0$, the value of ϕ in (11.12) which corresponds to the nonnegative solution $\{x_n\}_{n=-k}^{\infty}$ of Eq(11.11).

(c) Determine the set of all nonnegative initial conditions $x_{-k}, \ldots, x_0$ for which the corresponding solution $\{x_n\}_{n=-k}^{\infty}$ of Eq(11.11) converges to a period-$(k+1)$ solution of the form (11.12) with a given $\phi \geq 0$?

(See [6].)

Conjecture 11.4.17 *Let k be a positive integer. Show that every positive solution of*

$$x_{n+1} = \frac{1 + x_n + x_{n-k}}{x_{n-k+1}}, \quad n = 0, 1, \ldots$$

converges to a periodic solution of period $2k$.

Appendix A

Global Attractivity for Higher Order Equations

In this appendix we present some global attractivity results for higher order difference equations which are analogous to Theorems 1.4.5-1.4.8. We state the results for third order equations from which higher order extensions will become clear.

Let I be some interval of real numbers and let

$$f : I \times I \times I \to I$$

be a continuously differentiable function.

Then for every set of initial conditions $x_0, x_{-1}, x_{-2} \in I$, the difference equation

$$x_{n+1} = f(x_n, x_{n-1}, x_{n-2}), \quad n = 0, 1, \ldots \tag{A.1}$$

has a unique solution $\{x_n\}_{n=-2}^{\infty}$.

As we have seen in previous chapters, very often a good strategy for obtaining global attractivity results for Eq(A.1) is to work in the regions where the function $f(x, y, z)$ is monotonic in its arguments. In this regard there are eight possible scenarios depending on whether $f(x, y, z)$ is non decreasing in all three arguments, or non increasing in two arguments and non decreasing in the third, or non increasing in one and non decreasing in the other two, or non decreasing in all three.

Theorem A.0.1 *Let $[a, b]$ be an interval of real numbers and assume that*

$$f : [a, b]^3 \to [a, b]$$

is a continuous function satisfying the following properties:

(a) $f(x, y, z)$ is non-decreasing in each of its arguments;

(b) The equation

$$f(x, x, x) = x,$$

has a unique solution in the interval $[a, b]$.

Then Eq(A.1) has a unique equilibrium $\overline{x} \in [a, b]$ and every solution of Eq(A.1) converges to $\overline{x}$.

Proof. Set

$$m_0 = a \quad \text{and} \quad M_0 = b$$

and for $i = 1, 2, \dots$ set

$$M_i = f(M_{i-1}, M_{i-1}, M_{i-1}) \quad \text{and} \quad m_i = f(m_{i-1}, m_{i-1}, m_{i-1}).$$

Now observe that for each $i \geq 0$,

$$m_0 \leq m_1 \leq \ldots \leq m_i \leq \ldots \leq M_i \leq \ldots \leq M_1 \leq M_0,$$

and

$$m_i \leq x_k \leq M_i \quad \text{for } k \geq 3i + 1.$$

Set

$$m = \lim_{i\to\infty} m_i \quad \text{and} \quad M = \lim_{i\to\infty} M_i.$$

Then

$$M \geq \limsup_{i\to\infty} x_i \geq \liminf_{i\to\infty} x_i \geq m \tag{A.2}$$

and by the continuity of f,

$$m = f(m, m, m) \quad \text{and} \quad M = f(M, M, M).$$

In view of *(b)*,

$$m = M = \overline{x},$$

from which the result follows. □

Theorem A.0.2 *Let $[a, b]$ be an interval of real numbers and assume that*

$$f : [a, b]^3 \to [a, b]$$

is a continuous function satisfying the following properties:

(a) $f(x, y, z)$ is non-decreasing in x and $y \in [a, b]$ for each $z \in [a, b]$, and is non-increasing in $z \in [a, b]$ for each x and $y \in [a, b]$;

(b) If $(m, M) \in [a, b] \times [a, b]$ is a solution of the system

$$m = f(m, m, M) \quad \textit{and} \quad M = f(M, M, m)$$

then

$$m = M.$$

Then Eq(A.1) has a unique equilibrium $\overline{x} \in [a, b]$ and every solution of Eq(A.1) converges to $\overline{x}$.

Proof. Set

$$m_0 = a \quad \text{and} \quad M_0 = b$$

and for each $i = 1, 2, \ldots$ set

$$M_i = f(M_{i-1}, M_{i-1}, m_{i-1})$$

and

$$m_i = f(m_{i-1}, m_{i-1}, M_{i-1}).$$

Now observe that for each $i \geq 0$

$$a = m_0 \leq m_1 \leq \ldots \leq m_i \leq \ldots M_i \leq \ldots M_1 \leq M_0 = b.$$

and

$$m_i \leq x_k \leq M_i \quad \text{for} \quad k \geq 3i + 1.$$

Set

$$m = \lim_{i\to\infty} m_i \quad \text{and} \quad M = \lim_{i\to\infty} M_i.$$

Then (A.2) holds and by the continuity of f,

$$m = f(m, m, M) \quad \text{and} \quad M = f(M, M, m).$$

Therefore in view of *(b)*

$$m = M$$

from which the result follows. □

Theorem A.0.3 *Let $[a, b]$ be an interval of real numbers and assume that*

$$f : [a, b]^3 \to [a, b]$$

is a continuous function satisfying the following properties:

(a) $f(x, y, z)$ is non-decreasing in x and z in $[a, b]$ for each $z \in [a, b]$, and is non-increasing in $y \in [a, b]$ for each x and z in $[a, b]$;

(b) If $(m, M) \in [a, b] \times [a, b]$ is a solution of the system

$$M = f(M, m, M) \quad \textit{and} \quad m = f(m, M, m)$$

then

$$m = M.$$

Then Eq(A.1) has a unique equilibrium $\overline{x} \in [a, b]$ and every solution of Eq(A.1) converges to $\overline{x}$.

Proof. Set

$$m_0 = a \quad \text{and} \quad M_0 = b$$

and for each $i = 1, 2, \ldots$ set

$$M_i = f\left(M_{i-1}, m_{i-1}, M_{i-1}\right)$$

and

$$m_i = f\left(m_{i-1}, M_{i-1}, m_{i-1}\right).$$

Now observe that for each $i \geq 0$

$$a = m_0 \leq m_1 \leq \ldots \leq m_i \leq \ldots M_i \leq \ldots M_1 \leq M_0 = b.$$

and

$$m_i \leq x_k \leq M_i \quad \text{for} \quad k \geq 3i + 1.$$

Set

$$m = \lim_{i\to\infty} m_i \quad \text{and} \quad M = \lim_{i\to\infty} M_i.$$

Then (A.2) holds and by the continuity of f,

$$m = f\left(m, m, M\right) \quad \text{and} \quad M = f\left(M, M, m\right).$$

Therefore in view of *(b)*

$$m = M$$

from which the results follows. □

Theorem A.0.4 *Let $[a, b]$ be an interval of real numbers and assume that*

$$f : [a, b]^3 \to [a, b]$$

is a continuous function satisfying the following properties:

(a) $f(x, y, z)$ is non-increasing in x for each y and $z \in [a, b]$ and is non-decreasing in y and z for each $x, y \in [a, b]$;

(b) If $(m, M) \in [a, b] \times [a, b]$ is a solution of the system

$$M = f(m, M, M) \quad \textit{and} \quad m = f(M, m, m)$$

then

$$m = M.$$

Then Eq(A.1) has a unique equilibrium $\overline{x} \in [a, b]$ and every solution of Eq(A.1) converges to $\overline{x}$.

Proof. Set

$$m_0 = a \quad \text{and} \quad M_0 = b$$

and for each $i = 1, 2, \dots$ set

$$M_i = f(m_{i-1}, M_{i-1}, M_{i-1})$$

and

$$m_i = f(M_{i-1}, m_{i-1}, m_{i-1}).$$

Now observe that for each $i \geq 0$

$$m_0 \leq m_1 \leq \dots \leq m_i \leq \dots M_i \leq \dots M_1 \leq M_0 = b$$

and

$$m_i \leq x_k \leq M_i \quad \text{for} \quad k \geq 3i + 1.$$

Set

$$m = \lim_{i \to \infty} m_i \quad \text{and} \quad M = \lim_{i \to \infty} M_i.$$

Then (A.2) holds and by the continuity of f,

$$M = f(m, M, M) \quad \text{and} \quad m = f(M, m, m).$$

Therefore in view of *(b)*

$$m = M$$

from which the results follows. □

Theorem A.0.5 *Let $[a, b]$ be an interval of real numbers and assume that*

$$f : [a, b]^3 \to [a, b]$$

is a continuous function satisfying the following properties:

(a) $f(x, y, z)$ is non-decreasing in x for each y and $z \in [a, b]$ and is non-increasing in y and z for each $x \in [a, b]$;

(b) If $(m, M) \in [a, b] \times [a, b]$ is a solution of the system

$$M = f(M, m, m) \quad \textit{and} \quad m = f(m, M, M)$$

then

$$m = M.$$

Then Eq(A.1) has a unique equilibrium $\overline{x} \in [a, b]$ and every solution of Eq(A.1) converges to $\overline{x}$.

Proof. Set

$$m_0 = a \quad \text{and} \quad M_0 = b$$

and for each $i = 1, 2, \ldots$ set

$$M_i = f(M_{i-1}, m_{i-1}, m_{i-1})$$

and

$$m_i = f(m_{i-1}, M_{i-1}, M_{i-1}).$$

Now observe that for each $i \geq 0$

$$m_0 \leq m_1 \leq \ldots \leq m_i \leq \ldots M_i \leq \ldots M_1 \leq M_0 = b$$

and

$$m_i \leq x_k \leq M_i \quad \text{for} \quad k \geq 3i + 1.$$

Set

$$m = \lim_{i\to\infty} m_i \quad \text{and} \quad M = \lim_{i\to\infty} M_i.$$

Then (A.2) holds and by the continuity of f,

$$M = f(M, m, m) \quad \text{and} \quad m = f(m, M, M).$$

Therefore in view of *(b)*

$$m = M$$

from which the results follows. □

Theorem A.0.6 *Let $[a, b]$ be an interval of real numbers and assume that*

$$f : [a, b]^3 \to [a, b]$$

is a continuous function satisfying the following properties:

(a) $f(x, y, z)$ is non-decreasing in y for each x and $z \in [a, b]$ and is non-increasing in x and z for each $y \in [a, b]$;

(b) If $(m, M) \in [a, b] \times [a, b]$ is a solution of the system

$$M = f(m, M, m) \quad \text{and} \quad m = f(M, m, M)$$

then

$$m = M.$$

Then Eq(A.1) has a unique equilibrium $\overline{x} \in [a, b]$ and every solution of Eq(A.1) converges to $\overline{x}$.

Proof. Set

$$m_0 = a \quad \text{and} \quad M_0 = b$$

and for each $i = 1, 2, \dots$ set

$$M_i = f(m_{i-1}, M_{i-1}, m_{i-1})$$

and

$$m_i = f(M_{i-1}, m_{i-1}, M_{i-1}).$$

Now observe that for each $i \geq 0$

$$m_0 \leq m_1 \leq \dots \leq m_i \leq \dots M_i \leq \dots M_1 \leq M_0 = b$$

and

$$m_i \leq x_k \leq M_i \quad \text{for} \quad k \geq 3i + 1.$$

Set

$$m = \lim_{i\to\infty} m_i \quad \text{and} \quad M = \lim_{i\to\infty} M_i.$$

Then (A.2) holds and by the continuity of f,

$$M = f(m, M, m) \quad \text{and} \quad m = f(M, m, M).$$

Therefore in view of *(b)*

$$m = M$$

from which the results follows. □

Theorem A.0.7 *Let $[a, b]$ be an interval of real numbers and assume that*

$$f : [a, b]^3 \to [a, b]$$

is a continuous function satisfying the following properties:

(a) $f(x, y, z)$ is non-decreasing in z for each x and y in $[a, b]$ and is non-increasing in x and y for each $z \in [a, b]$;

(b) If $(m, M) \in [a, b] \times [a, b]$ is a solution of the system

$$M = f(m, m, M) \quad \textit{and} \quad m = f(M, M, m)$$

then

$$m = M.$$

Then Eq(A.1) has a unique equilibrium $\overline{x} \in [a, b]$ and every solution of Eq(A.1) converges to $\overline{x}$.

Proof. Set

$$m_0 = a \quad \text{and} \quad M_0 = b$$

and for each $i = 1, 2, \ldots$ set

$$M_i = f(m_{i-1}, m_{i-1}, M_{i-1})$$

and

$$m_i = f(M_{i-1}, M_{i-1}, m_{i-1}).$$

Now observe that for each $i \geq 0$

$$m_0 \leq m_1 \leq \ldots \leq m_i \leq \ldots M_i \leq \ldots M_1 \leq M_0 = b$$

and

$$m_i \leq x_k \leq M_i \quad \text{for} \quad k \geq 3i + 1.$$

Set

$$m = \lim_{i \to \infty} m_i \quad \text{and} \quad M = \lim_{i \to \infty} M_i.$$

Then (A.2) holds and by the continuity of f,

$$M = f(m, m, M) \quad \text{and} \quad m = f(M, M, m).$$

Therefore in view of *(b)*

$$m = M$$

from which the results follows. □

Theorem A.0.8 *Let $[a, b]$ be an interval of real numbers and assume that*

$$f : [a, b]^3 \to [a, b]$$

is a continuous function satisfying the following properties:
(a) $f(x, y, z)$ is non-increasing in all three variables x, y, z in $[a, b]$;
(b) If $(m, M) \in [a, b] \times [a, b]$ is a solution of the system

$$M = f(m, m, m) \quad \textit{and} \quad m = f(M, M, M)$$

then

$$m = M.$$

Then Eq(A.1) has a unique equilibrium $\overline{x} \in [a, b]$ and every solution of Eq(A.1) converges to $\overline{x}$.

Proof. Set

$$m_0 = a \quad \text{and} \quad M_0 = b$$

and for each $i = 1, 2, \ldots$ set

$$M_i = f(m_{i-1}, m_{i-1}, m_{i-1})$$

and

$$m_i = f(M_{i-1}, M_{i-1}, M_{i-1}).$$

Now observe that for each $i \geq 0$

$$m_0 \leq m_1 \leq \ldots \leq m_i \leq \ldots M_i \leq \ldots M_1 \leq M_0 = b$$

and

$$m_i \leq x_k \leq M_i \quad \text{for} \quad k \geq 3i + 1.$$

Set

$$m = \lim_{i\to\infty} m_i \quad \text{and} \quad M = \lim_{i\to\infty} M_i.$$

Then (A.2) holds and by the continuity of f,

$$M = f(m, m, m) \quad \text{and} \quad m = f(M, M, M).$$

Therefore in view of *(b)*

$$m = M$$

from which the results follows. □

The above theorems can be easily extended to the general difference equation of order p

$$x_{n+1} = f(x_n, x_{n-1}, \ldots, x_{n-p+1}), \quad n = 0, 1, \ldots \tag{A.3}$$

where

$$f : I^p \to I$$

is a continuous function. In this case, there are 2^p possibilities for the function f to be non decreasing in some arguments and non increasing in the remaining arguments. This gives 2^p results of the type found in Theorems A.0.1-A.0.8. Let us formulate one of these results.

Theorem A.0.9 *Let $[a, b]$ be an interval of real numbers and assume that*

$$f : [a, b]^p \to [a, b]$$

is a continuous function satisfying the following properties:

(a) $f(u_1, u_2, \ldots, u_p)$ is non decreasing in the second and the third variables u_2 and u_3 and is non increasing in all other variables.

(b) If $(m, M) \in [a, b] \times [a, b]$ is a solution of the system

$$M = f(m, M, M, m, \ldots, m) \quad \text{and} \quad m = f(M, m, m, M, \ldots, M)$$

then

$$m = M.$$

Then Eq(A.3) has a unique equilibrium $\overline{x} \in [a, b]$ and every solution of Eq(A.3) converges to $\overline{x}$.

Bibliography

[1] R. Agarwal, *Difference Equations and Inequalities. Theory, Methods and Applications*, Marcel Dekker Inc., New York, 1992.

[2] L. S. Allen, M. K. Hannigan, and M. J. Strauss, Mathematical analysis of a model for a plant-herbivore system, *Bulletin of Mathematical Biology* 55 (1993), 847-864.

[3] L. S. Allen, M. J. Strauss, M. J. Thorvilson, and W. N. Lipe, A Preliminary mathematical model of the Apple Twig Borer (Coleoptera: Bostrichidae) and Grapes on the Texas High Plains *Ecological Modelling* 58 (1991), 369-382.

[4] K.T. Alligood, T. Sauer and J.A. Yorke , *CHAOS An Introduction to Dynamical Systems, Springer-Verlag*, New York, Berlin, Heidelberg, Tokyo, 1997.

[5] American Mathemathical Monthly, Problem 10670, 106(1999), 967-968.

[6] A. M. Amleh and G. Ladas, Convergence to periodic solutions, *J. Differ. Equations Appl.* 7(2001), (to appear).

[7] A. M. Amleh, D. A. Georgiou, E. A. Grove, and G. Ladas, On the recursive sequence $x_{n+1} = \alpha + \frac{x_{n-1}}{x_n}$, *J. Math. Anal. Appl.* 233(1999), 790-798.

[8] A. M. Amleh, N. Kruse, and G. Ladas, On a class of difference equations with strong negative feedback, *J. Differ. Equations Appl.* 5(6)(1999), 497-515.

[9] A. M. Amleh, G. Ladas, and J. T. Hoag, Boundedness of solutions of a plant-herbivore system, *Proceedings of the Third International Conference on Difference Equations and Applications*, September 1-5, 1977, Taipei, Taiwan, Gordon and Breach Science Publishers, 1999, 1-5.

[10] R. Abu-Saris, On the periodicity of the difference equation $x_{n+1} = \alpha|x_n| + \beta x_{n-1}$, *J. Differ. Equations Appl.* 5(1)(1999), 57-69.

[11] E. Barbeau, B. Gelbard, and S. Tanny, Periodicities of solutions of the generalized Lyness recursion, *J. Differ. Equations Appl.* 1(3)(1995), 291-306.

[12] B. Beckermann and J. Wimp, Some dynamically trivial mappings with applications to the improvement of simple iteration, *Comput. Math. Appl.* 24(1998), 89-97.

[13] E. Camouzis, R. DeVault, and G. Ladas, On the recursive sequence $x_{n+1} = -1 + \frac{x_{n-1}}{x_n}$, *J. Differ. Equations Appl.* 7(2001), (to appear).

[14] K. L. Cooke, D. F. Calef, and E. V. Level, Stability or chaos in discrete epidemic models, *Nonlinear Systems and Applications*, V. Lakshmikantham, editor, 73-93, Academic Press, New York, 1977.

[15] K. Cunningham, M. R. S. Kulenović, G. Ladas, and S. Valicenti, On the recursive Sequence $x_{n+1} = \frac{\alpha+\beta x_n}{Bx_n+Cx_{n-1}}$, Proceedings of WCNA2000, Catania, Italy, July 19-26, 2000, *Nonlin. Anal. TMA*,(to appear).

[16] R. Devaney, *A First Course in Chaotic Dynamical Systems: Theory and Experiment*, Addison-Wesley, Reading, MA, 1992.

[17] R. Devaney, A Piecewise linear model for the zones of instability of an area-preserving map, *Physica D* 10(1984), 387-393.

[18] R. DeVault, L. Galminas, E. J. Janowski and G. Ladas, On the recursive sequence $x_{n+1} = \frac{A}{x_n} + \frac{B}{x_{n-1}}$, *J. Differ. Equations Appl.* 6(1)(2000), 121-125.

[19] R. DeVault, G. Ladas, and S. W. Schultz, On the Recursive Sequence $x_{n+1} = \frac{A}{x_n} + \frac{1}{x_{n-2}}$, *Proc. Amer. Math. Soc.* 126 (1998), 3257-3261.

[20] R. DeVault, G. Ladas, and S. W. Schultz, On the recursive sequence $x_{n+1} = \frac{A}{x_n x_{n-1}} + \frac{1}{x_{n-3}x_{n-4}}$, *J. Differ. Equations Appl.* 6(4)(2000), 481-483.

[21] S. Elaydi, *An Introduction to Difference Equations*, 2nd ed., Springer-Verlag, New York, 1999.

[22] S. Elaydi, *Discrete Chaos*, CRC Press, Boca Raton, FL, 2000.

[23] H. El-Metwally, E. A. Grove, and G. Ladas, A global convergence result with applications to periodic solutions, *J. Math. Anal. Appl.*, 245 (2000), 161-170.

[24] H. El-Metwally, E. A. Grove, G. Ladas, R. Levins, and M. Radin, On the difference equation $x_{n+1} = \alpha + \beta x_{n-1}e^{-x_n}$, Proceedings of WCNA2000, Catania, Italy, July 19-26, 2000, (to appear).

[25] H. El-Metwally, E. A. Grove, G. Ladas, and H. D. Voulov, On the global attractivity and the periodic character of some difference equations, *J. Differ. Equations Appl.* 7(2001), (to appear).

[26] J. Feuer, E. J. Janowski, and G. Ladas, Lyness-type equations in the third quadrant, *Nonlinear Anal.* 30(1997), 1183-1189.

[27] J. E. Franke, J. T. Hoag, and G. Ladas, Global attractivity and convergence to a two-cycle in a difference equation, *J. Differ. Equations Appl.* 5(2)(1999), 203-209.

[28] C. H. Gibbons, M. R. S. Kulenović, and G. Ladas, On the recursive sequence $x_{n+1} = \frac{\alpha+\beta x_{n-1}}{\gamma+x_n}$, *Math. Sci. Res. Hot-Line*, 4(2)(2000), 1-11.

[29] C. H. Gibbons, M. R. S. Kulenović, G. Ladas, and H. D. Voulov, On the trichotomy character of $x_{n+1} = \frac{\alpha+\beta x_n+\gamma x_{n-1}}{A+x_n}$, *J. Differ. Equations Appl.* 7(2001) (to appear).

[30] E. A. Grove, E. J. Janowski, C. M. Kent, and G. Ladas, On the rational recursive sequence $x_{n+1} = (\alpha x_n+\beta)/((\gamma x_n+C)x_{n-1})$, *Comm. Appl. Nonlinear Anal.* 1(1994), 61-72.

[31] E. A. Grove, C. M. Kent, R. Levins, G. Ladas, and S. Valicenti, Global stability in some population models , *Proceedings of the Fourth International Conference on Difference Equations and Applications*, August 27-31, 1998, Poznan, Poland, Gordon and Breach Science Publishers, 2000, 149-176.

[32] E. A. Grove, G. Ladas, L.C. McGrath, and C. T. Teixeira, Existence and behavior of solutions of a rational system, *Comm. Appl. Nonlinear Anal.* 8(2001), 1-25.

[33] J. Guckenheimer and P. Holmes, *Nonlinear Oscillations, Dynamical Systems, and Bifurcations of Vector Fields*, Springer-Verlag, New York, Berlin, Heidelberg, Tokyo, 1983.

[34] J. Hale and H. Kocak, *Dynamics and Bifurcations*, Springer-Verlag, New York, Berlin, Heidelberg, Tokyo, 1991.

[35] M. L. J. Hautus and T.S. Bolis, Solution to Problem E2721, *Amer. Math. Monthly*, 86(1979), 865-866.

[36] J. H. Jaroma , On the global asymptotic stability of $x_{n+1} = \frac{a+bx_n}{A+x_{n-1}}$, *Proceedings of the First International Conference on Difference Equations and Applications*, May 25-28, 1994, San Antonio, Texas, Gordon and Breach Science Publishers, 1995, 281-294.

[37] G. Karakostas, A discrete semi-flow in the space of sequences and study of convergence of sequences defined by difference equations, *M. E. Greek Math. Society*, (1989), 66-74.

[38] G. Karakostas, Convergence of a difference equation via the full limiting sequences method, *Differential Equations Dynam. Systems* 1(1993), 289-294.

[39] G. Karakostas, Asymptotic 2-periodic difference equations with diagonally self-invertible responses, *J. Differ. Equations Appl.* 6(3)(2000), 329-335.

[40] G. Karakostas, Ch. G. Philos, and Y. G. Sficas, The dynamics of some discrete population models, *Nonlinear Anal.* 17(1991), 1069-1084.

[41] W. G. Kelley and A. C. Peterson, *Difference Equations*, Academic Press, New York, 1991.

[42] V. L. Kocic and G. Ladas, *Global Behavior of Nonlinear Difference Equations of Higher Order with Applications*, Kluwer Academic Publishers, Dordrecht, Holland, 1993.

[43] V. L. Kocic, G. Ladas, and I. W. Rodrigues, On the rational recursive sequences, *J. Math. Anal. Appl.* 173 (1993), 127-157.

[44] V. L. Kocic, G. Ladas, G. Tzanetopoulos, and E. Thomas, On the stability of Lyness' equation, *Dynamics of Continuous, Discrete and Impulsive Systems* 1(1995), 245-254.

[45] W. A. Kosmala, M. R. S. Kulenović, G. Ladas, and C. T. Teixeira, On the recursive sequence $y_{n+1} = \frac{p+y_{n-1}}{qy_n+y_{n-1}}$, *J. Math. Anal. Appl.* 251(2000), 571-586.

[46] U. Krause, Stability trichotomy, path stability, and relative stability for positive nonlinear difference equations of higher order, *J. Differ. Equations Appl.* 1(4) (1995), 323-346.

[47] N. Kruse and T. Nesemann, Global asymptotic stability for discrete dynamical systems, *J. Math. Anal. Appl.* 235 (1999), 151-158.

[48] Y. K. Kuang and J. M. Cushing, Global stability in a nonlinear difference-delay equation model of flour beetle population growth, *J. Differ. Equations Appl.* 2(1)(1996), 31-37.

[49] M. R. S. Kulenović, Invariants and related Liapunov functions for difference equations, *Appl. Math. Lett.* 13 (7)(2000), 1-8.

[50] M. R. S. Kulenović and G. Ladas, Open problems and conjectures: On period two solutions of $x_{n+1} = \frac{\alpha+\beta x_n+\gamma x_{n-1}}{A+Bx_n+Cx_{n-1}}$, *J. Differ. Equations Appl.* 6(5) (2000), 641-646.

[51] M. R. S. Kulenović, G. Ladas, L. F. Martins, and I. W. Rodrigues, On the dynamics of $x_{n+1} = \frac{\alpha+\beta x_n}{A+Bx_n+Cx_{n-1}}$ facts and conjectures, *Comput. Math. Appl.*, (to appear).

[52] M. R. S. Kulenović, G. Ladas, and N. R. Prokup, On the recursive sequence $x_{n+1} = \frac{\alpha x_n+\beta x_{n-1}}{1+x_n}$, *J. Differ. Equations Appl.* 6(5) (2000), 563-576.

[53] M. R. S. Kulenović, G. Ladas, and N. R. Prokup, On a rational difference equation, *Comput. Math. Appl.* 41(2001), 671-678.

[54] M. R. S. Kulenović, G. Ladas, and W. S. Sizer, On the recursive sequence $x_{n+1} = \frac{\alpha x_n+\beta x_{n-1}}{\gamma x_n+Cx_{n-1}}$, *Math. Sci. Res. Hot-Line* 2(1998), no.5, 1-16.

[55] S. Kuruklis and G. Ladas, Oscillation and global attractivity in a discrete delay logistic model, *Quart. Appl. Math.* 50 (1992), 227-233.

[56] G. Ladas, Invariants for generalized Lyness equations, *J. Differ. Equations Appl.* 1(2)(1995), 209-214.

[57] G. Ladas, Open problems and conjectures, *J. Differ. Equations Appl.* 1(1)(1995), 95-97.

[58] G. Ladas, Open problems and conjectures, *J. Differ. Equations Appl.* 1(3)(1995), 317-321.

[59] G. Ladas, Open Problems and Conjectures, *J. Differ. Equations Appl.* 1(4)(1995), 413-419.

[60] G. Ladas, On the recursive sequence $x_{n+1} = \sum_{i=0}^{k} \frac{A_i}{x_{n-2i}}$, *J. Differ. Equations Appl.* 4(3)(1998), 311-313.

[61] G. Ladas, Open problems and conjectures, *J. Differ. Equations Appl.* 4(5)(1998), 497-499.

[62] G. Ladas, Open problems and conjectures, *J. Differ. Equations Appl.* 7(2001), (to appear).

[63] M. Martelli, *Introduction to Discrete Dynamical Systems and Chaos*, John Wiley & Sons, New York, 1999.

[64] Mathemathics Magazine, Problem 1572, 73(2000), 160-161.

[65] Ch. G. Philos, I. K. Purnaras, and Y. G. Sficas, Global attractivity in a nonlinear difference equation, *Appl. Math. Comput.* 62(1994), 249-258.

[66] E. C. Pielou, *An Introduction to Mathematical Ecology*, John Wiley & Sons, New York, 1965.

[67] E. C. Pielou, *Population and Community Ecology*, Gordon and Breach, New York, 1974.

[68] C. Robinson, *Stability, Symbolic Dynamics, and Chaos*, CRC Press, Boca Raton, FL, 1995.

[69] H. Sedaghat, Existence of solutions of certain singular difference equations, *J. Differ. Equations Appl.* 6(5)(2000), 535-561.

[70] H. Sedaghat, Geometric stability conditions for higher order difference equations, *J. Math. Anal. Appl.* 224(1998), 255-272.

[71] W. S. Sizer, Some periodic solutions of the Lyness equation , *Proceedings of the Fifth International Conference on Difference Equations and Applications*, January 3-7, 2000, Temuco, Chile, Gordon and Breach Science Publishers, 2001, (to appear).

[72] E. C. Zeeman, Geometric unfolding of a difference equation, http://www.math.utsa.edu/ecz/gu.html

[73] Y. Zheng, Existence of nonperiodic solutions of the Lyness equation $x_{n+1} = \frac{\alpha+x_n}{x_{n-1}}$, *J. Math. Anal. Appl.* 209(1997), 94-102.

[74] Y. Zheng, On periodic cycles of the Lyness equation, *Differential Equations Dynam. Systems* 6(1998), 319-324.

[75] Y. Zheng, Periodic solutions with the same period of the recursion $x_{n+1} = \frac{\alpha+\beta x_n+\gamma x_{n-1}}{A+Bx_n+Cx_{n-1}}$, *Differential Equations Dynam. Systems* 5(1997), 51-58.

Index